Springer Tracts in Modern Physics
Volume 136

Springer
Berlin
Heidelberg
New York
Barcelona
Budapest
Hong Kong
London
Milan
Paris
Santa Clara
Singapore
Tokyo

Springer Tracts in Modern Physics

Volumes 120–136 are listed at the end of the book

Covering reviews with emphasis on the fields of Elementary Particle Physics, Solid-State Physics, Complex Systems, and Fundamental Astrophysics

Manuscripts for publication should be addressed to the editor mainly responsible for the field concerned:

Gerhard Höhler
Institut für Theoretische Teilchenphysik
Universität Karlsruhe
Postfach 6980
D-76128 Karlsruhe
Germany
Fax: +49 (7 21) 37 07 26
Phone: +49 (7 21) 6 08 33 75
Email: hoehler@fphvax.physik.uni-karlsruhe.de

Johann Kühn
Institut für Theoretische Teilchenphysik
Universität K arlsruhe
Postfach 6980
D-76128 Karlsruhe
Germany
Fax: +49 (7 21) 37 07 26
Phone: +49 (7 21) 6 08 33 72
Email: johann.kuehn@physik.uni-karlsruhe.de

Thomas Müller
IEKP
Fakultät für Physik
Universität Karlsruhe
Postfach 6980
D-76128 Karlsruhe
Germany
Fax:+49 (7 21) 6 07 26 21
Phone: +49 (7 21) 66 08 35 24
Email:mullerth@vxcern.cern.ch

Roberto Peccei
Department of Physics
University of California, Los Angeles
405 Hilgard Avenue
Los Angeles, California 90024-1547
USA
Fax: +1 310 825 9368
Phone: +1 310 825 1042
Email: robertop@lands.sscnet.ucla.edu

Frank Steiner
Abteilung für Theoretische Physik
Universität Ulm
Albert-Einstein-Allee 11
D-89069 Ulm
Germany
Fax: +49 (7 31) 5 02 29 24
Phone: +49 (7 31) 5 02 29 10
Email: Steiner@physik,uni-ulm.de

Joachim Trümper
Max-Planck-Institut
für Extraterrestrische Physik
Postfach 1603
D-85740 Garching
Germany
Fax: +49 (89) 32 99 35 69
Phone: +49 (89) 32 99 38 81
Email: jtrumper@mpe-garching.mpg.de

Peter Wölfle
Institut für Theorie
der Kondensierten Materie
Universität Karlsruhe
Kaiserstraße 12
D-76131 Karlsruhe
Germany
Fax: +49 (7 21) 69 81 50
Phone: +49 (7 21) 6 08 35 90-33 67
Email: woelfle@tkm.physik.uni-karlsruhe.de

Ulrike Woggon

Optical Properties of Semiconductor Quantum Dots

With 126 Figures

 Springer

Dr. Ulrike Woggon

Institut für Angewandte Physik
Universität Karlsruhe
Postfach 6980
D-76128 Karlsruhe

Library of Congress Cataloging-in-Publication Data

Woggon, Ulrike, 1958-
    Optical properties of semiconductor quantum dots / Ulrike Woggon.
       p.   cm. -- (Springer tracts in modern physics, ISSN 0081-3869
    ; v. 136)
    Includes bibliographical references and index.
    ISBN 3-540-60906-7 (Hardcover : alk. paper)
    1. Wide gap semiconductors--Optical properties. 2. Quantum
  electronics.   I. Title. II. Series: Springer tracts in modern
  physics ; 136.
  QC1.S797 vol. 136
  [QC611.8.W53]
  539 s--dc20
  [537.6'22]   96-44746
                                                             CIP

Physics and Astronomy Classification Scheme (PACS):
72.40+w, 78.66.Db, 78.66.Fd, 81.05.Cy, 81.05.Ea

ISSN 0081-3869
ISBN 3-540-60906-7 Springer-Verlag Berlin Heidelberg New York

© Springer-Verlag Berlin Heidelberg 1997
Printed in Germany

Typesetting: Camera-ready copy by the author using a Springer TeX macro-package
SPIN: 10515691   56/3144-5 4 3 2 1 0 – Printed on acid-free paper

Preface

Systematic research on quantum dots began in the early 1980s with the identification of quantum confinement in small, nanocrystalline semiconductor inclusions in glasses and colloids, though the use of semiconductor-doped glasses as edge-filters or as Q-switches in lasers is much older. The "quantum dot" has been rapidly developed to an intensively investigated model system of basic research, extending the physics of reduced dimensions to all three space coordinates. A variety of theoretical calculations have been published concerning, for example, the determination of energy states and wave functions. At that time, comparison with experimental results was limited by the problem of how to manufacture quantum dots with reproducible properties. It looked like a failing attempt to confirm the emerging theory of three-dimensional spherical confinement by reliable experimental results. As an efficient tool to monitor the electronic properties of quantum dots, methods of linear and nonlinear optics have been successfully established in experiments. Thus, the first detailed information could be gained by optical spectroscopy in the studies of the development from clusters to the solid state with growing dimensions of the crystallites.

This work is aimed at pointing out to the reader the current knowledge about the optical properties of quantum dots. Through discussions of experiments into linear and nonlinear optics and into electro-optics, the intrinsic electronic properties of three-dimensionally confined semiconductors will be illustrated. In choosing the topics of the chapters of this book, I tried to touch on all important experimental activities in the field of quantum dots.

The book starts with the illustration of the growth and precipitation process of the crystallites in different environments, including epitaxial growth. To characterize the intrinsic electronic properties we compare theoretical calculations of the confined energy levels with experimentally identified energy positions found by using the great variety of spectroscopic methods available. The development from one-electron–hole-pair states to two-pair states and many-particle systems will follow. I will deal with some corrections due to deviations from the ideal particle-in-a-spherical-box problem, introduced, for example, by peculiarities in the band structure, the finite height of the potential barriers, or by differences in the dielectric constants between the semiconductor and the host material. Since knowledge of not only the line positions but also the line width is necessary to understand the optical spectra exhibited by the quantum dots, we will study the different dephasing mechanisms in quantum dots. The analysis of phase relaxation (and energy relaxation) is still at its beginning and I will give a few first examples. However, this field is going to be a central topic of future work. Furthermore, I will look at physical properties that are basic for possible applications. The physical origin of nonlinear and electro-optic properties will be discussed in the context of their exploitation in device concepts. Competitive mechanisms that lower the nonlinear or electro-optic response will be analyzed. The advantage that could be obtained by using matrix materials other than glass will be discussed. The review covers all semiconductor materials suited for creating zero-dimensional structures, starting from the wide-gap I–VII and II–VI compounds, and going via the group-IV materials and III–V compounds through the whole visible spectrum up to the near-infrared spectral range.

In spite of the work done during the last decade our current knowledge about quantum dots is rather fragmentary. The research on semiconductor materials of lower dimensions has finally left its infancy but is still far from being mature. For the next few years this field of science promises many new problems. In any case, the "quantum dot" is a model system in which we can study subjects from many different fields of science and no single semiconductor bulk system allows us to involve such a great diversity of knowledge.

In writing this book, I have benefitted from numerous discussions with many colleagues and students. Special thanks go to C. Klingshirn for his continuous interest and support, and for many valuable discussions. Many of the results could be only obtained due to close and fruitful collaborations, and I am indebted to A. Uhrig, M. Saleh, O. Wind, H. Giessen, W. Langbein, M. Portuné, V. Sperling, F. Gindele, A. Lohde, and H. Spöcker for their contributions. I am also very grateful to many colleagues for encouraging discussions and suggestions. It is a pleasure for me to thank L. Banyai, M.G. Bawendi, Al.L. Efros, A.I. Ekimov, B. Fluegel, D. Fröhlich, S.V. Gaponenko, O. Gogolin, B. Hönerlage, H. Kalt, S.W. Koch, M. Müller, S. Nomura, N. Peyghambarian, U. Rössler, I. Rückmann, L. Spanhel, and E. Tsitsishvili. Direct financial support for this work has been given by the Deutsche Forschungsgemeinschaft.

Karlsruhe, September 1996   *Ulrike Woggon*

Contents

1. Introduction

Present semiconductor physics appears to be the physics of systems of reduced dimensionality. Artificially made semiconductor structures show a surprising variety of new interesting properties that are completely different from solid-state bulk materials and have never been observed there. The fabrication of single or periodic potential wells by simply combining two semiconductor materials of different bandgap energies and with spatial dimensions confining the motion of electrons and holes, results in many impressive possibilities for engineering of the semiconductor properties. Two-dimensionally layered material systems exhibiting quantum confinement in the direction of the growth axis are widely investigated and their study represents a new, rapidly developing field in solid-state physics.

It is understandable that scientists' efforts are directed to further decrease the dimensions to quasi-one-dimensional or zero-dimensional structures. One possibility for obtaining zero-dimensional structures is the inclusion of spherical semiconductor particles in a dielectric, transparent matrix. Such a structure is mesoscopic in all three dimensions, i.e. its radius is large compared to the lattice constant, but comparable to the spatial extension of the wave functions of excitons, electrons or holes in the corresponding bulk semiconductor material. To define these particles many different terms have been used such as quantum dots (QD), nanocrystals, microcrystallites (MC), Q-particles, or nanoclusters. The small semiconductor spheres of, for example, II–VI or I–VII compounds can be grown in different matrices, such as glasses, solutions, polymers or even cavities of zeoliths, and by different manufacturing processes, for example by melting and annealing processes, by organometallic chemistry or by sol-gel techniques. Evidence for quasi zero-dimensional structures has likewise been obtained by investigating the epitaxial growth on highly mismatched substrates. It results in the development of small islands on the substrate surface with high regularity and sufficiently small sizes to show quantum confinement.

Quantum dots are very attractive and interesting objects for scientific research of three-dimensionally confined systems. Their investigation requires an insight into many different branches of knowledge such as solid-state physics, molecular physics, photochemistry, nonlinear optics and ultrafast spectroscopy, materials sciences and structural analysis. The simplest, naturally

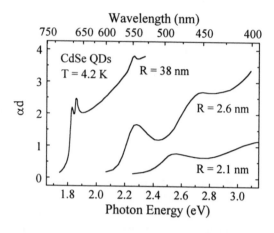

Fig. 1.1. Spectra of linear absorption of CdSe nanocrystals with different radii $R = 2.1$ nm, 2.6 nm and 38 nm, embedded in a borosilicate glass matrix, data taken from Ekimov et al. (1985a, 1993)

given zero-dimensional systems are spherical semiconductor nanocrystals embedded in glasses or organic matrices. This has proved to be the ideal model system for the study of basic questions of three-dimensional confinement in semiconductors. Beyond basic research there is always the question of possible applications. Promising ideas exist regarding the fields of integrated optics (active elements in waveguide structures, fast switching devices, light-emitting and laser diodes) and of physical chemistry (solar energy conversion, photocatalysis).

The growth of quantum dots in glasses is one of the oldest and most frequently used techniques. The first hints of the existence of small inclusions of, for example, CdSe and CdS in silicate glasses causing its yellow to red color were connected to the development of X-ray analysis and published in the early 1930s by Rocksby (1932). Since the 1960s, semiconductor doped glasses have been widely applied as sharp-cut color filter glasses in optics. The concept of quantum confinement, and by that the distinction between the coloring of the glasses by *changes in stoichiometry* of CdS_xSe_{1-x} mixed crystals or by *size changes* of the binary nanocrystals, was introduced by Efros and Efros (1982), and confirmed experimentally by Ekimov and Onushenko (1984). At the same time, the change of color of semiconductor colloidal solutions has been discussed in the context of quantum confinement effects by Rosetti et al. (1984). A period of control of the growth process of these nanocrystalline semiconductors followed in the 1980s combined with a detailed investigation of their linear and nonlinear optical behavior.

As demonstrated in Fig. 1.1, a first approach in understanding the behavior of quantum dots is mainly the investigation of their optical properties, in particular of their absorption spectra. Using the data first reported by Ekimov et al. (1985a, 1993) the size-dependent change of the absorption spectra is plotted in Fig. 1.1 for CdSe-doped glasses. Nanocrystals of large radii (e.g. $R = 38$ nm) show the typical spectrum of bulk CdSe. The spectrum is characterized by the sharp band-edge and, close to that, by the series of exciton states. Three peaks are located at the energy of the A-exciton ($E = 1.826\,\mathrm{eV}$), B-exciton ($E = 1.85\,\mathrm{eV}$) and C-exciton transitions [($E = 2.26\,\mathrm{eV}$, $T = 1.8\,\mathrm{K}$, (Landolt–Börstein 1982)] originating from the spin–orbit and crystal-field splitting of the valence-band states. The spectral positions of the absorption peaks shift to higher energies and the lines become broader with decreasing sizes of the nanocrystals. The explanation for these spectral changes in the absorption spectra by a *size*-dependent effect was the beginning of the intensive research on three-dimensional quantum confinement in solid-state semiconductor composite materials.

The main intention of this book is an approach to the three-dimensional quantum confinement in semiconductors by analyzing the optical properties of the corresponding semiconductor structures. To do this, one has to consider the absorption coefficient of an ensemble of quantum dots inside a transparent matrix. The averaged absorption spectrum $\bar{\alpha}(\omega)$ can then be expressed by

$$\bar{\alpha}(\omega) = \frac{p}{\bar{V}_{\mathrm{QD}}} \int \mathrm{d}R \; \frac{4\pi}{3} \; R^3 \; P(R) \; \alpha_{\mathrm{QD}}(\omega, R) \tag{1.1}$$

with p the volume fraction of the semiconductor material, $\bar{V}_{\mathrm{QD}}$ the average quantum dot volume, R the radius and $P(R)$ a characteristic distribution function for the dot sizes, as well as $\alpha_{\mathrm{QD}}(\omega, R)$ the absorption coefficient of a single quantum dot. As a result of the quantum confinement effect, the absorption coefficient α_{QD} is strongly dependent on the radius R of the dot. The absorption spectrum is given by a series of Lorentzian lines for the ground and excited states at energies $E_{\mathrm{QD}}^j = \hbar\omega_j$, with homogeneous line widths Γ_j and oscillator strengths f_j

$$\alpha_{\mathrm{QD}}(\omega, R) \sim \sum_j f_j(R) \; \frac{\dfrac{\hbar\,\Gamma_j(R)}{2}}{(E_{\mathrm{QD}}^j(R) - \hbar\omega)^2 + \left(\dfrac{\hbar\,\Gamma_j(R)}{2}\right)^2} \; . \tag{1.2}$$

Equations (1.1) and (1.2) show that one needs information about the radius R, the size distribution $P(R)$, and the semiconductor volume fraction p to correlate the appearance of structures in the absorption spectrum to, for example, electronic states of the quantum dots, as well as suitable relations for the size dependence of $E(R)$, $\Gamma(R)$ and $f(R)$, the energy, homogeneous line broadening and oscillator strength, respectively. When the quantum dots are additionally exposed to, for example, electric fields, pressure or high photon densities, further interesting problems appear since then quantum confinement and external forces combine and influence the energy, line broadening and oscillator strength of the optical transitions.

The size-dependence of the optical properties of quantum dots has been one of the main subjects of research work over the last decade. The enormous growth of experimental data (e.g. Ekimov et al. 1985a, 1991; Borelli et al. 1987; Brus 1991; Bawendi et al. 1990a; Henglein 1988; Wang 1991a), and the increasing exactness of the theoretical concepts (e.g. Banyai and Koch 1993; Haug and Banyai 1989; D'Andrea et al. 1992) resulted in a steadily growing understanding of the electronic and optical properties of quantum dots. In the present book an attempt has been made to provide an overview of the variety of phenomena which can influence the physical properties of quantum dots, starting with the matrix material, size, structure, and interfaces, but considering also excitation densities, external and internal fields, lattice properties etc. The choice of topics covered here is determined by experimental points of view. However, this work is being written in a period of rapid development and therefore without the claim of presenting a complete discussion of all experiments carried out in this field during the last few years.

We will mainly concentrate on quantum dots of II–VI semiconductors embedded in glass matrices, but consider also other semiconductor compounds as well as other types of matrix materials. The widespread quantum dot systems based on II–VI materials show absorption structures in the visible and near-ultraviolet part of the spectrum and are therefore compatible with a great number of laser sources used for experiments. The experimental techniques applied comprise almost all standard experiments of linear and nonlinear optics such as absorption, steady-state and time-resolved luminescence, pump-and-probe spectroscopy, and degenerate and non-degenerate four wave mixing. In this work I start with the growth process of the nanocrystals and then consider under which conditions a nanocrystal can be defined as a quantum dot. I will demonstrate that the application of experimental methods taken from nonlinear optics of bulk materials is a powerful tool for identifying energy states and homogeneous line broadening. As a main topic, the new, specific properties related to excitons and biexcitons in three-dimensionally confined systems will be discussed. I will deal with the different mechanisms of phase relaxation and their influence on three-dimensional confinement. The interaction with the matrix, the specific aspect of optics of composite materials and the role of interfaces will be further topics. Furthermore, I will give a short overview of experimental results which have been obtained when applying external fields. Possible applications will be discussed as well.

Currently, quantum dots derived from III–V compounds are also being studied intensively. They are prepared by etching techniques of two-dimensionally confined layered structures, ion implantation or island-like epitaxial growth. The detailed analysis of the properties of these structures is beyond the scope of this work. However, a brief survey will be given accompanied by a list of corresponding references. Recent summaries of the development in the field of interesting transport properties of quantum dots, such as single electron transport and Coulomb blockade, can be found, for example, in Kuchar et al. (1990), Merkt (1990), Reed (1993), Geerligs et al. (1993) and in the references therein.

Over the last three years enormous progress could be observed in the investigation of quantum dots obtained from indirect-gap semiconductor materials, for example Si-nanoclusters. Although we cannot provide a comprehensive presentation of this field, a short summary of the first studies will be given, taking into account that indirect-gap quantum dots are of some specific interest. For more information I refer, for example, to Littau et al. (1993), Takagahara and Takeda (1992), Brus (1994) and references therein.

In the field of theory the influence of Coulomb interaction, the description of the electron and hole states, the electron-phonon coupling, the problem of the dielectric confinement and of the interface, high-density phenomena, field action etc. have been treated during the last five years. A comprehensive representation of the theory of semiconductor quantum dots may be found in the book of Banyai and Koch (1993). Therefore a detailed review of the results of theory of quantum dots has been omitted. I refer to theory only in the case where the theoretical results are complemented or confirmed by the presented experiments.

Finally, I should mention that the concept of quantum dots is already included in modern textbooks (Peyghambarian et al. 1993; Haug and Koch 1993; Klingshirn 1995). For further information about quantum confinement, the book by Bastard (1988) provides a very good comparison of the physics of two-dimensionally confined layered systems and three-dimensionally confined quantum dots.

2. Growth of Nanocrystals

This chapter deals with the problems of the growth process of nanocrystals, i.e. with the analysis of the laws of growth, the final sizes of the nanocrystals and expected size distributions. Different procedures have been reported for the growth of nanocrystals, such as the growth inside the cavities of a zeolith, or the growth in an organic environment by stabilization of the nanocrystals by organo-metallic ligand molecules at the surface (Chestnoy et al. 1986; Henglein 1988; Wang et al. 1989, 1995; Bawendi et al. 1990a; Bagnall and Zarzycki 1990; Spanhel and Anderson 1991). In glasses, it is common practice to describe the growth of nanocrystals by the model of condensation from a supersaturated solid solution (Lifshitz and Slezov 1961; Ekimov et al. 1985a, 1991).

The understanding of the growth process is a prerequisite for the understanding of all basic properties of three-dimensional confinement in semiconductor doped glasses. Often, the description of quantum dots starts with the introduction of different classification schemes, for example the characterization of the confinement range with respect to the ratio of the radius R of the nanocrystal to the Bohr radius a_B of the exciton in the corresponding bulk material (strong, medium or weak confinement). Further classification is possible with respect to the treatment within the frame of the effective mass approximation or within the concepts of cluster physics. Other interesting parameters are the volume-to-surface ratio, the ratio between the confinement energy and the phonon energies, and the differences in the dielectric constants between the semiconductor and the matrix material.

Thus, the growth and the following characterization procedure have to provide a minimum of data before the study of basic electronic and optical properties of quantum dots becomes meaningful. The most important information to be obtained from growth analysis concerns the sizes, the size distribution, stoichiometry, structure and the interface configuration of the nanocrystals.

2.1 Growth of Nanocrystals in Glass Matrices

2.1.1 The Diffusion-Controlled Growth Process

How big will nanocrystals be if they are grown in a certain matrix, over a fixed time and within a given range of temperature? How much time is needed to achieve a certain mean radius and what size distribution is then obtained? To describe the growth process it is interesting to look for models giving answers to these questions.

In the course of a real growth process, however, several growth stages may occur and sometimes even coexist. For example, we know the nucleation process with the formation of stable nuclei, the normal growth stage where these nuclei grow from the supersaturated solution and, at low values of supersaturation, the onset of the competitive growth where the larger particles grow due to the dissolution of smaller particles. For simplification, theoretical description commonly starts with the separate analysis of these different stages of growth to obtain simple analytical expressions.

The attempts to find a functional description of the growth process, or generally of phase transitions, go back to the early 1930s (Becker and Döring 1935). The main idea of the classical homogeneous nucleation theory consists in the ansatz that the phase transition proceeds (at constant temperature) between a supersaturated vapor of n monomers (molecules) and droplets containing in $(i \geq 2)$ monomers. The starting point is the rate equation determining the number of monomers n inside the droplets (clusters). This equation relates the gain g_n and the losses l_n, where the size of the clusters changes by the gain or the loss of one monomer

$$\frac{\mathrm{d}n}{\mathrm{d}t} = g_n - l_n \ . \tag{2.1}$$

The resulting change in the monomer density $\mathrm{d}n/\mathrm{d}t$ is determined by the difference between the number of monomers impinging on and evaporating from the cluster through the surface into the supersaturated vapor phase. The variety of the occurring growth mechanisms is expressed by introducing the explicit terms for g_n and l_n. For instance, for the growth of thin films the gain g_n can be determined by surface diffusion, interface transfer (nonthermic transport), or cluster movement and unification (Turnbull 1956; Chakraverty 1967; Abraham 1974; Landau and Lifshitz 1974; Koch 1984). The losses l_n were derived by analyzing the energy balance between surface and volume energy contributions of the condensed phase.

The primary aim of growth analysis is to evaluate asymptotic functions for the cluster size distribution. Involving detailed expressions for g_n and l_n in (2.1) and combining this equation with the matter conservation law and the equation of continuity, the cluster growth laws have been calculated. One of the most famous models is the Lifshitz–Slezov model (Lifshitz and Slezov 1961). It is often used to fit the growth process of semiconductor nanocrystals in glasses and is therefore presented here.

A search is underway for expressions for $f(R,t)$, the size distribution for clusters with radius R at the time t. The continuity equation of the probability density in cluster space coordinates is

$$\frac{\partial f(R^3,t)}{\partial t} + \frac{\partial}{\partial R^3}[f(R^3,t)\cdot \dot{R}^3] = 0 . \qquad (2.2)$$

$\dot{R} = \mathrm{d}R/\mathrm{d}t$ is the cluster growth velocity and has to be determined from the rate equation (2.1). This can be done by the transformation of the average number of monomers per cluster to the average cluster size (R) and the consideration of the conservation of the total amount of matter. The solution of (2.2), and thus $f(R^3,t)$, depends on the different mechanisms considered for g_n and l_n and leads to different expressions for $\dot{R}$.

The Lifshitz–Slezov model assumes that the transport of monomers to the cluster surface is realized by diffusion in a supersaturated solution. The monomer distribution in time and space is obtained from the diffusion equation which yields for $\dot{R}$

$$\frac{\mathrm{d}R}{\mathrm{d}t} = \frac{R}{D}[\Delta - \frac{\alpha}{R}] \qquad (2.3)$$

with D the diffusion coefficient, Δ the supersaturation

$$\Delta = \bar{C} - C_\infty \qquad (2.4)$$

and

$$\alpha = \frac{2\sigma}{k_{\mathrm{B}}T}\, v\, C_\infty . \qquad (2.5)$$

C_∞ is the concentration of the saturated solution in equilibrium, $\bar{C}$ the averaged concentration of the supersaturated solution, v the volume of one monomer in the solution and σ the surface energy (see for more details Lifshitz and Slezov 1961; Abraham 1974; Koch 1984). The critical droplet radius is defined by

$$R_{\mathrm{cr}}^0 = \frac{\alpha}{\Delta_0} . \qquad (2.6)$$

The supersaturation is a function of time. Thus, for every value Δ of the supersaturation there exists a critical cluster radius $R_{\mathrm{cr}} = R_{\mathrm{cr}}(t)$, for which the cluster is in equilibrium with the solute. If $R > R_{\mathrm{cr}}$, the cluster will grow; in the case of $R < R_{\mathrm{cr}}$ it will dissolve.

For illustration, a scheme of the growth process of a supersaturated solution is given in Fig. 2.1. A small cluster is instable because the concentration around it is too high, and a large cluster is instable because there is a very low concentration and a depletion region around it. The monomers feel a concentration gradient driving the diffusion of matter from smaller to larger clusters. A diffusion-controlled competitive growth of the nanoclusters is the result. In the framework of the Lifshitz–Slezov analysis, the asymptotically

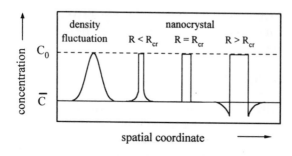

Fig. 2.1. Concentration profile around the nanocluster during the growth process. $\bar{C}$ is the average concentration within the supersaturated solution, C_0 is the concentration inside the nanocluster

stable size is the one for which the concentration directly near the surface C_R is equal to the average concentration of the matrix ($C_R = \bar{C}$).

Applying the Lifshitz–Slezov model to the growth process in glassy matrices, the surface free energy, one of the parameters determining R_{cr}, will depend on both matrix environment and semiconductor material. Furthermore, in the Lifshitz–Slezov analysis the number of nuclei is considered to be constant. To avoid a high formation rate of nuclei, a low value of supersaturation is required. When we carry out the asymptotic analysis we find that the final size distribution evaluated by Lifshitz and Slezov (1961) is asymmetrical (see below, Fig. 2.3) and independent of the initial distribution. Its maximum value does not correspond to the critical radius and is slightly shifted to larger radii $R > R_{cr}$. The slope to larger sizes is abrupt and nanoclusters of sizes larger than twice the critical radius do not exist. However, the function gradually decreases toward smaller sizes with a long tail.

The volume of the cluster grows linearly in time corresponding to

$$R_{\mathrm{av}} = \left(\frac{4\alpha D}{9}t\right)^{1/3} . \tag{2.7}$$

It depends on the diffusion coefficient D and, via α (2.5), on temperature and surface energy.

The asymptotic limit of the classical Lifshitz–Slezov analysis describes the growth process in the case of long heat treatment times. But what determines the initial radii and the growth process at very early stages?

Let us consider the thermodynamics of nucleation without the problem of long-time kinetics. In the framework of steady-state homogeneous nucleation theory, condensation is driven [in analogy to the simple problem of liquid droplets in water vapor (Abraham 1974; Turnbull 1956)] by the condition of minimization of free energy. In equilibrium the number of condensed droplets N can be estimated by an Arrhenius ansatz

$$N = N_0 \exp(-\Delta F/k_{\mathrm{B}}T) . \tag{2.8}$$

ΔF is the free energy necessary for the formation of the nuclei. Thus, to become stable, a cluster must acquire an excess of free energy compared to the single vapor molecules. In the vapor/droplet system the free energy is given by

$$\Delta F = 4\pi R^2 \sigma - \frac{4\pi R^3}{3v} k_B T \ln \frac{p}{p_\infty} . \tag{2.9}$$

The first term is the contribution of the surface free energy, the second term represents the contribution to ΔF caused by the bulk free energy change, where p/p_∞ is the vapor supersaturation. If R is the droplet radius, the positive term varies in proportion to R^2, while the opposing negative term varies in proportion to R^3. Thus, regardless of the magnitude of the coefficients of these terms, in some sufficiently small interval of R above zero the positive R^2 term dominates the R^3 term.

We see that the number of nuclei is strongly influenced by the value of the supersaturation. For any case $p/p_\infty > 1$ the free energy ΔF exhibits a maximum. The radius at which ΔF attains its maximum is obtained from the condition

$$\frac{\mathrm{d}\,\Delta F}{\mathrm{d}R} = 0 \tag{2.10}$$

and again corresponds to the critical nucleus radius

$$R_{cr} = \frac{2\sigma\,v}{k_B T \ln \frac{p}{p_\infty}} . \tag{2.11}$$

The larger the supersaturation, the smaller the critical radius R_{cr} and even small clusters can grow. Low values of supersaturation lead to a large critical cluster radius R_{cr}. Using (2.11) and (2.9), the height of the energy barrier for the nucleation ΔF_{cr} can be evaluated as

$$\Delta F_{cr} = \frac{16\pi\sigma^3 v^2}{3\,(k_B T \ln \frac{p}{p_\infty})^2} = \frac{4\pi}{3} R_{cr}^2\,\sigma . \tag{2.12}$$

When the degree of supersaturation is increased gradually, the number of impinging monomers grows and the probability increases that more and more monomers will overcome the top of the activation barrier in a given time. Additionally, the increase of supersaturation reduces the height of the energy barrier to such an extent that the probability that some of the subcritical clusters will grow accidentally to supercritical size nearly approaches unity. At that value of supersaturation, homogeneous nucleation becomes an effective process and the phase transition sets in. Nucleation is intrinsically a probabilistic event. The radius fluctuation probability around the critical radius has to be described by statistics. In this simple consideration, based on homogeneous nucleation theory, one would expect a Gaussian symmetric size distribution around R_{cr}.

This model from the system water/vapor can be transferred to more general cases by introducing the expression Δ_0 for an universal supersaturation. Also nanocrystals in glassy matrices can be grown near nucleation to obtain

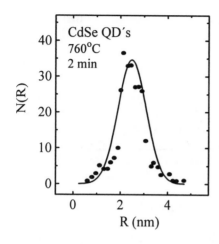

Fig. 2.2. Size distribution of CdSe nanocrystals in the early stages of growth. Data from Liu and Risbud (1990). The solid line corresponds to a Gaussian fit

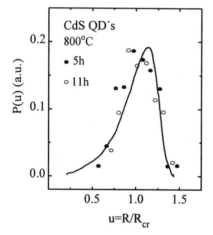

Fig. 2.3. Size distribution of CdS nanocrystals after long-time heat treatment. Data from Potter and Simmons (1988). The solid line corresponds to the asymptotic distribution function derived by Lifshitz and Slezov (1961)

very small sizes (see 2.1.2). The number of nanoclusters and their starting radius can be controlled by the supersaturation p/p_∞, the surface free energy σ (glass matrix properties) and the temperature T. In the growth process actually realized the various stages often coexist, i.e. the nucleation and normal growth process may exist simultaneously. In this case the number of nuclei is no longer constant. Calculations made by Shepilov (1992) examining this kind of growth behavior resulted in the interesting fact that a time interval can also exist for which the size distribution changes while the mean radius of the crystallites remains constant.

Experimentally, the asymptotic growth behavior has been studied much better than any other precipitation stage since just this stage is used for commercial color filter glasses (Tsekhomskii 1978; Golubkov et al. 1981; Uhrig et al. 1992). Some examples of the different stages of growth of nanocrystals

are shown in Figs. 2.2 and 2.3. A proposal for the precipitation of nanocrystals in the very early time range has been made by Liu and Risbud (1990).
One example of a growth procedure over a long time period has been given by Potter and Simmons (1988). The two representative growth regimes
have been compared with respect to the size distribution realized. Figure 2.2
shows the sizes of CdSe nanocrystals and their distribution after a very short
heat treatment of 2 minutes at the relatively high temperature of 760 °C.
The size distribution has been determined experimentally and is shown to be
symmetrical. It has been fitted by a Gaussian function,

$$P(R) = (2\pi)^{-1/2} \, \Delta R^{-1} \, \exp\left(-\frac{(R - R_{\mathrm{av}})^2}{2(\Delta R)^2}\right) \tag{2.13}$$

also plotted in the figure. The average size obtained by TEM measurements
is $R \approx 2.5\,\mathrm{nm}$. The standard deviation of $\Delta R \approx 1.4\,\mathrm{nm}$ is still large. However,
the average distance between the nanocrystals is only 20 nm and thus a high
concentration of nuclei has been realized. At this early stage of growth the
sizes are predominantly determined by the matrix properties (surface energy
σ) and the concentration of cadmium and selen ions inside the matrix (supersaturation). Thus, very small crystals could be grown showing clear quantum
confinement.

A further idea for the control of mean size and size distribution is the
attempt to realize the asymptotic growth law. One example is shown in Fig.
2.3 (Potter and Simmons 1988). Here CdS nanocrystals have been grown at
800 °C over 5 and 11 hours with the aim of achieving the asymptotic limit
of growth. The size has been determined experimentally and compared with
the Lifshitz–Slezov size distribution function (Fig. 2.3)

$$P(u) = \begin{cases} 3^4 2^{-5/3} \; \mathrm{e}\, u^2 \, (u + 3)^{-7/3} \left(\frac{3}{2} - u\right)^{-11/3} \exp\left[\left(\frac{2u}{3} - 1\right)^{-1}\right] \\ \qquad\qquad \text{for } u < 1.5 \\ \qquad 0 \qquad \text{for } u > 1.5 \end{cases} \tag{2.14}$$

with $u = R/R_{\mathrm{cr}}$. The average radius of these large nanocrystals is around
18 nm, with a deviation of about 8 nm. As will be discussed later, these sizes
correspond to the weak confinement range.

In both cases the variation in size of about 50 % is still substantial and
represents a crucial problem for studying the intrinsic electronic properties.
During the last few years, much effort has been made to diminish the size
distribution, as will be outlined in more detail in the following sections.

2.1.2 Preparation of II–VI Nanocrystals

The range of semiconductor materials used for the preparation of nanocrystals covers almost all kinds of known substances absorbing in the ultraviolet,
visible and near infrared range of the spectrum. For the II–VI compounds
the realization of nanocrystals in glassy matrices has been reported for CdS,
CdSe, as well as the corresponding mixed crystals $CdS_x Se_{1-x}$ (e.g. Ekimov et

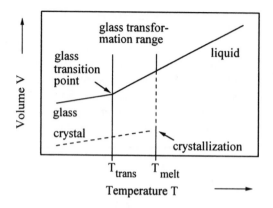

Fig. 2.4. Schematic drawing of the temperature dependence of the glass volume for the transition liquid–glass in comparison to the case of crystallization

al. 1985a, 1991; Borelli et al. 1987; Potter and Simmons 1988; Yan and Parker 1989; Roussignol et al. 1989; Liu and Risbud 1990; Woggon et al. 1991b; Rodden et al. 1994; Esch et al. 1992, Ricolleau et al. 1996), for CdTe (Esch et al. 1990, 1992), for $Cd_{1-x}Zn_xS$ (Yükselici et al. 1995) and for ZnO (Borelli et al. 1987). It is obvious that sol-gel preparation methods or the growth process in micelle media are completely different from the growth process of nanocrystals in glasses (see Sect. 2.2). Other preparation steps are applied in the production of an ordered quantum dot array by molecular beam epitaxy and nanolithography (see Sect. 2.5). Even for a similar composition of glassy matrices one would expect discrepancies in the growth behavior if differences between the melting temperature of the semiconductor and the glass transition temperature exist, as for example observed for I–VII compared to II–VI compounds. Owing to the diversity of preparation possibilities, the following section is restricted to the growth of CdS and CdSe nanocrystals in borosilicate glasses. Since this class of materials is representative and most widely investigated, this limitation seems to be justified.

The Glass Matrix. Glass is a solid amorphous material, which can also be considered as an undercooled melt. Its peculiarity is, as for all amorphous materials, that the melting point is not clearly defined and the transition from liquid to glass is continuous. During this transition, the glass transformation range, all physical properties change continuously, in particular the viscosity as the most important. The phenomenon of glass formation is illustrated in Fig. 2.4. At high temperatures the glass constituents are molten and in a liquid phase. With decreasing temperatures the volume of the liquid decreases accompanied by a simultaneous increase in the viscosity of the melt. At the melting point T_{melt}, the possibility of crystallization principally exists in connection with a strong change in the volume. Sometimes this transition can also be observed, in particular in the case of very slow cooling rates. However, if a rapid cooling rate is applied, no crystallization occurs and the glass is in the metastable thermodynamic state of an undercooled liquid. Further decreasing of the temperature increases the viscosity to such an extent that the molecules of the melt can no longer arrange themselves in an equilibrium state. In a viscosity range of $\sim 10^{12}$ Pas the glass is in solid configuration. The temperature that is attributed to this transition has been defined as the glass transition temperature T_g.

The strong temperature dependence of the viscosity of glass can be seen if comparing typical values measured in the different temperature regions. For $T = 1700$ K and $T = 1100$–1200 K, two temperatures in the liquid range, the viscosity of the glass compound is 1–10 Pas and 10^4 to 10^8 Pas, respectively. About 10^{12} Pas are typical for the glass transition range ($T_g \sim 600$ K depending on the glass constituents). Values of 10^{17} to 10^{19} Pas for the viscosity are obtained at room temperature connected with a high fragility of the glass material.

The glass constituents consist of two main components, the network former and the network modifier ions (see e.g. Scholze 1977; Neuroth 1987). The network formers are predominantly oxides, the most important of which are silicon dioxide SiO_2, aluminium oxide Al_2O_3 and boron trioxide B_2O_3. They determine the structural network of the amorphous solid glass. The use of *boro*silicate glasses instead of pure silicate glasses is connected with the wish to achieve a convenient range of melting temperatures. SiO_2 and B_2O_3 possess very different melting and glass transition temperatures [$T_{melt} = 1996$ K (SiO_2), $T_{melt} = 723$ K (B_2O_3), $T_g = 1500$ to 1900 K (Si_2O), $T_g = 470$ to 530 K (B_2O_3), from Scholze (1977)]. Therefore, a variation of the composition of the glass network formers also permits us to vary the melting and the glass transition temperatures. In the case of CuBr- and CuCl-doped glasses, for which the semiconductor components have very low melting points, the glass transition temperature can thus be chosen near the semiconductor melting point. For the II–VI materials with higher melting temperature this optimization is mostly less successful and considerable differences exist between the semiconductor melting point and these two temperatures, T_{melt} and T_g.

The modifier ions enhance the chemical stability of the glass. The best known and investigated is Na_2O, which is able to break bridges between the Si–O–Si bonds. The viscosity of the glass and thus the diffusion velocity of the ions is mainly influenced by the network modifier ions used.

The range of temperatures for the annealing process can be derived from this discussion of the basic properties of the glass matrix. It is the range between T_g and T_{melt}. The set of free parameters for heat treatment can be chosen between $500\,°C$ and 700–$800\,°C$ depending on the modifier ions and the composition of the glass forming components. In Table 2.1 some examples of glass compositions used are summarized together with the parameters of the following growth procedure of the semiconductor nanocrystals.

The glass matrix represents the confining barrier material for the small semiconductor inclusions. To enable us to estimate the height and shape of this potential barrier, the absorption spectrum of the undoped glass and of

Table 2.1. Comparison of base glass compositions and preparation methods for different CdS- and CdSe-doped glasses

II–VI Compound	Batch	Wt%	Melting Process	Heat Treatment	Ref.
CdSe, CdS	SiO_2 Na_2O B_2O_3 Al_2O_3 ZnO CdO Sb_2O_3 CdO Se (S)	68.9 5.6 11.5 1.0 11.3 0.59 1.1 0.67 (0.35) 0.14 (0.15)	1300– 1400 °C	575–750 °C 0.5–16 h one- and two-step annealing	Borelli et al. (1987)
CdS, CdSe	SiO_2 Na_2O B_2O_3 K_2O CaO (ZnO) Cd,S (Se)		1400 °C	548 °C 4–12 h 800 °C 0.1–11 h two-step annealing	Potter and Simmons (1988)
CdSe	SiO_2 Na_2O Al_2O_3 ZnO CdO Se, Sn	63.3 15.4 4.0 17.3	1400 °C 2 h	520 °C, 1 h 625–850 °C 0.5–100 h one- and two-step annealing	Yan and Parker (1989)
CdS, CdSe	SiO_2 K_2O B_2O_3 CaO BaO CdS, CdSe	56 24 8 3 9	1400 °C 1.5 h	577– 735 °C some hours	Liu and Risbud (1990)
CdSe	SiO_2 K_2O B_2O_3 ZnO CdSe	54 20 6 20	1250– 1400 °C 2 h	580–600 °C 10 h–10 d one-step annealing	Woggon et al. (1991a,b)
$CdS_{0.4}Se_{0.6}$	Si O C Zn Na K	16 53 18 5 1 7		923 °C 120 h	Goerigk et al. (1994)
CdS $Cd_{1-x}Zn_xS$	SiO_2 B_2O_3 Na_2O CaO K_2O ZnO	60 20 8 6 4 0.4	1100 °C	600–725 °C 0.25 h–15 d	Yükselici et al. (1995)

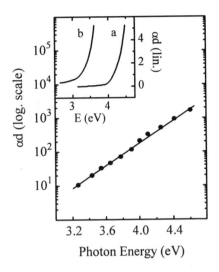

Fig. 2.5. Onset of the absorption for the glass matrix used, for example, by Woggon et al. (1991a,b). The inset shows in a linear scale the change in the absorption tail after adding the semiconductor components cadmium and selen to the melt (**b**) in comparison to the undoped base glass (**a**)

the doped glass before annealing are shown in Fig 2.5 for the glass composition used by Woggon et al. (1991a,b). It is difficult to define an energy gap for this class of materials from the onset of the absorption edge only, as usually has been done in semiconductors. In the undoped glass without cadmium and selen ions, an exponential absorption onset over more than three orders of magnitude is observed in the spectral range between 3 and 5 eV. From that fact, the barrier height can be estimated at around 1–2 eV assuming a gap of the semiconductor material of 2 eV. Although the exact partition of the potential between the conduction and valence band (the band offset) is not known, the assumption of infinitely high barriers is a justified approximation for many applications. However, after the addition of the semiconductor components to the glass constituents the semiconductor ions give rise to the development of a long absorption tail, as shown in the insert. This tail can reach the visible range of the spectrum. Then the infinitely high, vertical barrier is no longer a good approximation. Furthermore, the dissolved ions inside the glass matrix produce trap states, which can be populated directly by electrons that have escaped from the optically excited nanocrystal, as will be shown later. For the glass matrix presented in Fig. 2.5b this results in a broad luminescence band with a maximum around 600 nm (Woggon 1991b).

Heat Treatment, Sizes and Size Distribution. The models discussed in Sect. 2.1.1 are a good guide for finding an optimum growth regime with well-defined sizes and size distributions. The interesting questions are: what is the most suitable matrix composition and heat treatment? Does an additional preannealing step help to promote the nucleation and what are the consequences for the size distribution? What is the difference between a one- and a two-step annealing procedure? Is it possible to observe the Lifshitz–Slezov distribution within sizes where quantum confinement occurs?

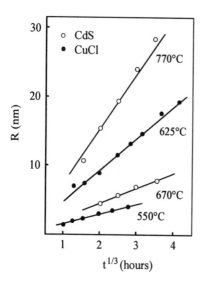

Fig. 2.6. Experimental proof of the $t^{1/3}$ dependence of the radii of CdS and CuCl nanocrystals (2.7) grown at different temperatures. From Ekimov et al. (1985a)

The first results reported in the literature aimed at the realization of the Lifshitz–Slezov distribution or at the experimental proof of this distribution. Clear evidence for a diffusion-controlled phase separation from a supersaturated solid solution has been obtained by Ekimov et al. (1985a). For the chosen matrix material the characteristic $t^{1/3}$ dependence of the nanocrystal radii (2.7) could already be obtained for small sizes of the crystallites (Fig. 2.6).

Likewise, experiments by Potter and Simmons (1988) partly demonstrated the diffusion-limited ripening model in the case of CdS crystallites (Figs. 2.3 and 2.7, Table 2.1). The CdS nanocrystals were produced by growing at high temperatures (> 700 °C) over long periods of time (several hours up to one day). The size distribution can roughly be approximated by the Lifshitz–Slezov function (2.14); however, the sizes are still large and the nanocrystals show a weak confinement effect only. The spectra exhibit well-defined structures in absorption although the Lifshitz–Slezov distribution has a relatively large deviation from the average size of the nanocrystals. The reason is that for sizes larger than the Bohr radius the average deviation in the radii is only weakly transformed into a change in energy of the optical transitions and therefore does not strongly influence the spectral shape.

Often a two-step heat treatment has been applied to create a high number of nuclei. In the first step the glass is heated for several hours at temperatures slightly above the glass transition temperature T_g prior to the second high-temperature annealing step. The difference in the temperatures is more than a few hundred degrees. Analyzing the two-step procedure, it has been found that a preannealing step, preceding the second treatment, increases only the number of formed nuclei, but no change in the final sizes and size distribution could be measured (Potter and Simmons 1988).

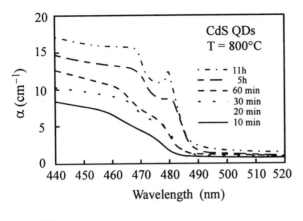

Fig. 2.7. Spectra of the linear absorption of CdS nanocrystals grown at high temperatures and over different times. Data taken from Potter and Simmons (1988)

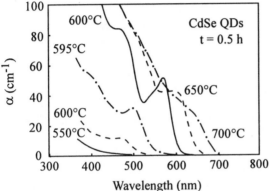

Fig. 2.8. Absorption spectra of CdSe nanocrystals for samples with a heat treatment of 0.5 h at various temperatures. Data taken from Borelli et al. (1987)

Similar results have been obtained by Yan and Parker (1989), confirming that the two-step procedure improves the size distribution only in a minor way. When the temperatures between the first and second heat treatment differ only by some tens of degrees, a small improvement could be observed by Gaponenko et al. (1993). The advantage of an additional nucleation phase, i.e. the increase of the number of nuclei created, is lost if the high temperature heat treatment exceeds temperatures of 650 °C. However, as will be discussed later (Sect. 5), a preannealing step can significantly influence the interface configuration and gives rise to changes in the defect-related scattering mechanisms and homogeneous line widths.

Growing the nanocrystals by realizing a diffusion-controlled precipitation (for example, by growing at high temperatures over long periods of time), one has the advantage of a well-known asymptotic distribution function independent from the starting conditions and fixed in its half width. The disadvantage is that the growth process carried out in the asymptotic limit does not result in very small nanocrystals.

Borelli et al. (1987) proposed that smaller crystallites can be grown with a much better size distribution if deviating from the asymptotic limit. For the matrix composition used (Table 2.1), the existence of the asymmetrical Lifshitz–Slezov distribution could be explicitly excluded for nanocrystals smaller than 5 nm radius, whereas it was observed for nanocrystals in commercial filter glasses with larger sizes. For the experimental glasses, the duration of the heat treatment has been chosen to be 0.5 hours (Fig. 2.8). During the growth, a clear tendency of increasing size distribution could be confirmed when heat treatment time was increased. For a heat treatment of 600 °C/0.5 h, a radius of 1.5±0.25 nm has been found (which corresponds to a deviation of 16 %); at 650 °C/0.5 h, the measured sizes are $R = 2.2 \pm 0.6$ nm (24 % deviation) and at 700 °C/0.5 h, the radii are 4±1.1 nm (27 % deviation), respectively.

Liu and Risbud (1990) and Yan and Parker (1989) attempted to study the first phase of the growth process at very early stages. CdSe nanocrystals were developed in different matrices at 800 °C after a very short treatment of 10 min (Yan and Parker 1989) or 760 °C and 2 min (Liu and Risbud 1990), respectively. The averaged radii obtained for the 800 °C/10 min and the 760 °C/2 min processes were 5 nm and 2.5 nm, respectively. The size distribution is symmetrical and well-fitted with a Gaussian, but the standard deviation lies between 20 % (Yan and Parker 1989) and 40 % (Liu and Risbud 1990) and is still substantial.

To summarize, a growth process carried out to achieve the asymptotic limit leads to the Lifshitz–Slezov distribution with large radii of nanocrystals. The size distribution is fixed in the limit of the Lifshitz–Slezov model and is relatively large. However, due to the large radii (i.e. only weak confinement), sometimes the inhomogeneous broadening of the spectra is small and well-defined peaks can be obtained. For specially selected matrix materials (see e.g. Ekimov et al. 1985a), the maximum size can be minimized to become as small as $R \sim 5$ nm, resulting in somewhat stronger confinement for semiconductor materials with larger Bohr radii (e.g. CdSe).

To grow small crystallites inside borosilicate glasses, it is necessary to find other growth procedures, for example, growth near nucleation. As already discussed, one possibility consists in the rapid cessation of the process after some minutes of growth at high temperatures. This procedure yields small radii and symmetrical distributions. However, it is sometimes likewise connected with a distribution of sizes of about 30 %, caused by high nucleation rates resulting from the high temperatures and large supersaturation. A further possibility to minimize the size deviation will be shown in the next section.

CdSe Grown near Nucleation. In Sect. 2.1.1 the thermodynamics of nucleation and the following diffusion-controlled growth process were discussed. The critical radius of nuclei and the rate of nucleation were found to be determined by the supersaturation and the surface free energy. In glasses these

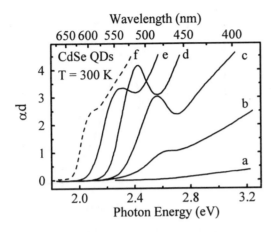

Fig. 2.9. CdSe nano-crystals grown in a one-step procedure near nucleation. (a) glass before heat treatment, (b) 580°C/10 h, (c) 580°C/20 h, (d) 580°C/80 h, (e) 600°C/170 h, (f) glass with different composition 700°C/8h. Data taken from Müller et al. (1992)

two parameters can be controlled by the semiconductor ion concentration and the matrix composition. Therefore, to obtain small nanocrystals, a one-step procedure at low supersaturation, realized by small ion concentrations in the matrix, has been carried out by Woggon et al. (1991a,b) and Müller et al. (1992). To prevent competitive growth between the nanocrystals, temperatures near the glass transformation point were used ensuring low diffusion coefficients. The development of nanocrystals is supposed to proceed only by thermodynamic nucleation followed by diffusion in the direct surrounding of the nanocluster and without interaction between them. The aim of these experiments is the investigation of the initial stage of diffusion-controlled growth near nucleation for the glass composition used and the determination of (i) the maximum radius attainable in this first stage, (ii) the narrowest size distribution, and (iii) the growth regime for the onset of and the transition to the stage of strong competitive growth and ripening. The batch became molten within a range of 1250–1400 °C, and was poured onto a copper plate to inhibit uncontrolled nucleation, and then the heat treatment was carried out at very low temperatures $T \leq 600$ °C. The corresponding absorption spectra are shown for selected samples in Fig. 2.9 measured at room temperature. Starting with the samples treated at 580 °C for 10 and 20 hours, well-defined peak structures, a shift to lower energy (longer wavelength), and an increase in the absolute absorption αd with increasing treatment time were found (Fig. 2.9a–c). The narrowest half width and highest αd-value were measured for a sample treated at 580 °C for 80 hours showing the absorption maximum at 510 nm (Fig 2.9d). A further increase in the heat treatment time at 580 °C did not result in a further measurable shift of the absorption peak. Therefore, the temperature was raised to 600 °C. The decreased viscosity of the glass matrix allowed a more effective diffusion and thus the growth rate rose again. After 170 hours a substantial shift of the absorption peak was observed and the absorption maximum was now found at 532 nm. However, there was no further increase in the absolute value of αd compared to the sample treated

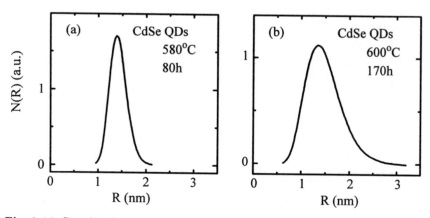

Fig. 2.10. Size distribution of the CdSe nanocrystals grown at 580°C/80 h (a) and 600°C/170 h (b) measured by small angle X-ray scattering. From Woggon et al. (1991a)

at 580 °C for 80 hours. For the matrix and supersaturation chosen, no further growth could be achieved by an increase in the growth time. Evidently, the dissolved semiconductor material was used up and the remaining volume concentration was too small (i.e. the distance between the nanocrystals was too large) and the temperature was too low to support the onset of the competitive growth. To obtain larger nanocrystals, a new growth procedure had to be started by changing the matrix composition, the ion concentration and by using higher temperatures (Fig. 2.9f). The smooth modulation of the absorption spectrum for this growth regime is also a clear hint at the development of wider size distributions.

To investigate the correlation between the conditions of heat treatment and the resulting size distributions, small angle X-ray scattering (SAXS) was applied with a detection limit of $R \leq 0.8$ nm of the setup used. In Fig. 2.10 the size distributions are plotted for the two samples treated at 580 °C for 80 h and 600 °C for 170 h. They were obtained from a numerical fitting procedure of the X-ray scattering intensity (Woggon et al. 1991a; Müller et al. 1992). Obviously the mean sizes of the CdSe nanocrystals are similar, but the particle size distribution of the sample treated at 600 °C has a larger mean deviation. Therefore we conclude that the samples treated at 580 °C represent the first stage of growth near the nucleation stage. The increase of αd shows the continuous conversion of Cd and Se ions dissolved in the glass matrix to nanocrystals of CdSe. At this stage, a symmetric size distribution was found with the maximum radius of 1.6 ± 0.2 nm, i.e. a spread of only 13 %. This value is one of the best reported for glasses [Roussignol et al. (1989) reported a value of 10 to 15 % for a commercial RG630 melt]. The size distribution of Fig. 2.10b and the absence of a further increase of the absorption maximum in the spectra indicate the transition to structural ripening for the sample grown at 600 °C. Yet, the regime at 600 °C is still at the beginning of the competitive diffusion-controlled growth and far from the asymptotic limit described by Lifshitz–Slezov. Consequently, for the growth procedure in glass the initial growth near nucleation should be one possibility to control and minimize the half width of the distribution of the radii. At present, further improvement has been attained by growing the nanocrystals in other matrix materials like polymers or by sol-gel technologies, where other mechanisms of growth and related laws exist and therefore also other possibilities for control (see e.g. Murray et al. 1993, and Sect. 2.2).

A very interesting development of sizes during the growth was recently found by Champagnon et al. (1993) confirming the above discussions. For a RG630-based CdS_xSe_{1-x}-doped glass (around SiO_2 46 %, K_2O 20 %, ZnO 21 %, TiO_2 6 %, B_2O_3 4 %) grown at 590, 620, and 645 °C over 2 days, the size distributions typical for the beginning and the asymptotic stage of growth were determined applying different size-analyzing methods, for example, high-resolution transmission electron microscopy, low-frequency inelastic Raman scattering and small angle X-ray scattering. The first sample of the CdS_xSe_{1-x}-doped glass grown at 590 °C for 2 days contains nanocrystals with average radii of 2.75 nm and a symmetric mean deviation of 0.75 nm. Growing a second sample with the heat treatment at 620 °C for 2 days results in somewhat larger nanocrystals. The size distribution, however, cannot be considered simply as a shifted or scaled version of the distribution of sample (1); it seems more likely that in addition to a Gaussian-distributed ensemble of nanocrystals around an average radius of 4 ± 1.2 nm, a second sharp peak from sizes around 6 nm appears. In a certain range of growth between the beginning and the asymptotic limit, the size distribution can develop two maxima; one is remaining from the growth near nucleation and the second is attributed to the average radius of the developing asymptotic distribution. In a third sample grown at 645 °C for 2 days, the Lifshitz–Slezov size-distribution function can already be recognized.

Nanocrystals grown near nucleation have the inherent characteristic of small size and large surface to volume ratio. It can be regarded as a peculiarity of these small nanocrystals that the surface configuration is not stable after finishing the growth procedure and, hence, can be further transformed by, for example, laser illumination. The energy provided by a strong UV-radiation source, here a nitrogen laser, is sufficient to create and change small clusters of the semiconductor material. In Fig. 2.11 the development of the maximum around 500 nm in the spectrum of glasses containing Cd and Se ions and small CdSe clusters, respectively, is a clear hint for a continuation of the growth procedure at room temperature by laser illumination for the chosen matrix composition and size of nanocrystals. As will be shown later, this sensitivity to laser illumination has to be taken into account carefully when discussing the nonlinear optical properties of very small nanocrystals.

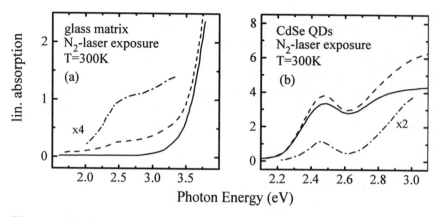

Fig. 2.11. (a) Absorption of the untreated glass before (solid line) and after (dashed line) laser exposure. (b) Absorption of very small CdSe nanocrystals before (solid line) and after (dashed line) annealing by laser exposure. The dashed-dotted lines are the difference spectra

2.1.3 Other Semiconductor Materials in Glass Matrices

In a similar way as reported above, nanocrystals of I–VII compounds such as CuCl, CuBr and CuI have been incorporated in glasses (Gilliot et al. 1989; Ekimov et al. 1985b; Gogolin et al. 1991; Woggon et al. 1994d; Itoh et al. 1994) as well as the material PbS in an alkali aluminosilicate glass by Borelli and Smith (1994). Some examples of the composition of the base glass chosen are given in Table 2.2.

The development of copper halide crystals in silica glasses exhibits a distinct difference from the II–VI materials, which will be outlined in more detail. Crystals of I–VII compounds have a very low melting temperature near $T_{melt} \approx 400\,°C$. The growth temperature, however, is determined by the matrix composition and ranges between 550 and 600 °C, similar to the one known from the growth of II–VI nanocrystals. Additionally, the relation $\alpha_{glass} < \alpha_{cryst}$ holds for the thermal expansion coefficients of the glass matrix and the I–VII semiconductor. In consequence, the crystallites are liquid during their growth and might appear, after cooling down to room temperature, inside a pore with a size somewhat larger than the crystallite size. Some hints of such behavior have been obtained from studies of hydrostatic pressure effects. The exciton bands in CuCl and CuBr quantum dots did not show any shift when hydrostatic pressure was applied under the experimental conditions reported by Kulinkin et al. (1988) and Vasil'ev et al. (1991).

Exploiting the relation $T_{melt} < T_{growth}$, Gaponenko et al. (1992) proposed a so-called 'fine annealing' process to improve the size distribution. Because T_{melt} is size-dependent a prolonged annealing process at T_{anneal} close to but less than T_{melt} can lead to sharper excitonic resonances.

Table 2.2. Comparison of base glass compositions and preparation
methods for CuCl- and CuBr-doped glasses

I–VII Compound	Batch	Wt%	Melting Process	Heat Treatment	Ref.
CuCl	SiO_2	60.34		600°C	Gilliot et al.
	Al_2O_3	9.5		0.5–12h	(1989)
	$Na_2B_4O_7$	29.02		two-step	
	CuCl	0.94		annealing	
	NaCl	0.1–2.0			
CuBr	SiO_2	48.8	1400 °C	550 –	Woggon et al.
	Na_2O	13.5		660°C	(1994)
	B_2O_3	28.3		1 h	
	Al_2O_3	9.4		two-step	
	CuO	0.3		annealing	
	CdO	0.5			
	SnO	0.1			
	CuBr				

CuI nanocrystals have been prepared by Gogolin et al. (1991). Here the
optical spectra clearly reflect the coexistence of the cubic and hexagonal
phase of the semiconductor inclusions.

It should be mentioned that copper halide nanocrystals likewise have been
found in alkali halide crystals heavily doped with copper (Itoh et al. 1988a;
Masumoto et al. 1993). Starting from NaCl, KBr, and KI monocrystals, crystallites of CuCl, CuBr, and CuI can be developed in this way. The preparation mostly starts from a Bridgman-type growth of the matrix single crystal
(e.g. NaCl) with an additional small admixture of highly purified CuCl powder. In addition to that technique, crystals of NaCl or KCl doped with CuCl
can be grown from the melt. Hence the crystallites grow by a sophisticated interplay between diffusion and coagulation of Cu^+ ions in the monocrystalline
NaCl or KCL matrix, and no functional dependence of the mean radius on
the heat treatment temperature can be expected. These structures show the
advantage of well-defined interfaces due to the chemical affinity of the host
and the embedded nanocrystals. Fröhlich and his colleagues (1995) succeeded
in determining the orientation of CuCl nanocrystals in NaCl matrix by using
two-photon absorption (TPA). From the polarization dependence of the TPA
signal it has been found that the crystal axis of the CuCl nanocrystals and
the NaCl matrix are parallel to each other.

To summarize, both crystallites of II–VI and I–VII compounds can be
developed by use of commercial glass technologies. The optical spectra show
quantum confinement effects but are very sensitive to the growth dynamics,
which determines the mean size and the size distribution. An important way
to obtain minimum deviations in the mean size of the crystallites is to perform
the growth process at the stage near nucleation over long periods of time.

2.2 Growth of Nanocrystals in Organic and Related Matrices

From the above discussion we can state that both fast nucleation and slow growth dynamics should be adjusted to obtain monodisperse nanocrystals. A further way to accomplish this fairly well is to grow the crystallites by chemical reactions in liquid or micelle media or in polymers. Here the matrix acts not only as the stabilizer; it also determines the chemical interface configuration and therefore controls the sizes via chemical equilibrium conditions. The growth of nanocrystals can be controlled by the choice of the solvent and the concentration of the reacting species as well as the reaction temperature and duration. Compared to glass technologies, the obvious advantage of these methods is the low preparation temperature (usually no more than 200 °C), and the realization of very narrow size distributions ($< 5\%$). In the following we will briefly review some of the methods; for more information see, for example, Alivisatos et al. (1988), Henglein (1988), Steigerwald and Brus (1989), Herron et al. (1990), Murray et al. (1993), Wang (1995) and references therein.

2.2.1 Chemical Preparation Methods

The most common preparation methods for II–VI nanocrystals in other than glass matrices are based on chemical replacement reactions between chemical compounds providing metal ions (Cd^{2+}, Zn^{2+} etc.) and those containing chalcogenide ions (S^{2-}, Se^{2-}, Te^{2-} etc.). The general reaction scheme, here for the example of CdSe, reads as

$$N(Cd^{2+} + Se^{2-}) \longrightarrow (CdSe)_N . \tag{2.14}$$

The synthesis most often proceeds in a two-step procedure. In a first step the cadmium salts or similar substances ($CdCl_2$, $Cd(OCOCH_3)_2$, $Cd(NO_3)_2$) are dissolved in organic solvents before, in a second step, the oppositely charged S^{2-} or Se^{2-} ions are added by the admixture of H_2S, H_2Se, Na_2S_x etc.

For the carrier of the chalcogenide ions, the use of organometallic reagents has proved to be very efficient (Steigerwald et al. 1988). In contrast to hydrogen compounds, the organometallic reagents are more stable and soluble in various organic solvents. Moreover, they are also able to carry capping molecule groups which can terminate the further growth process (see below). The compound $Si(CH_3)_3$ [Trimethylsilyl, (TMS)] is one of these organometallic groups that exists both in the form of $S(TMS)_2$, $Se(TMS)_2$ and $Te(TMS)_2$. Here, the driving force for the chemical reaction is the formation of strong covalent bonds.

One of the first attempts to produce II–VI nanocrystals was the manufacture of colloidal suspensions where, for example, CdS was formed by precipitation of Cd^{2+} in aqueous solution by adding H_2S. Additionally, a small

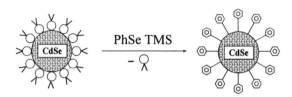

Fig. 2.12. Capping of CdSe nanocrystals with phenyl groups, according to Steigerwald et al. (1989), Ph – phenyl, Se – selen, TMS – trimethylsilyl

amount of natrium polyphosphate was added for stabilization. When carrying out the reaction with an excessive Cd^{2+} concentration, a surface passivation was observed by formation of a $Cd(OH)_2$ layer (Spanhel et al. 1987a,b). CdS crystallites could be grown down to very small sizes. The existence of very small stable CdS superclusters (55 atoms) could be demonstrated with astonishingly regular shape, like the small pyramids by Hilinski et al. (1988), Wang and Herron (1990), Herron et al. (1993), and Vossmeyer et al. (1995).

A very convenient way to control the growth is the use of microheterogeneous media like the reverse micelle media (Steigerwald and Brus 1989; Petit et al. 1990). A reverse micelle media is a microemulsion where a small amount of water in a hydrocarbon solvent is surrounded by surfactants (molecules having both hydrophobic and hydrophilic functional groups). A well-known micellar media is the AOT (0.12 M)/water(0.96 M) mixture [AOT = (bis(2-ethylhexyl)sulfosuccinate, disodium salt)]. In these micellar media the final particle size is governed by the molar ratio of the water to the surfactant. CdSe nanocrystals in reverse micelles are protected against agglomeration because of the surfactants. However, they can grow larger if either ionic or organometallic sources of Cd or Se atoms are added. The crystal surface can be chemically modified by replacement of the chalcogenide ion in the organometallic group by a phenyl group Ph (C_6H_5), as successfully applied by Bawendi et al. (1990a) and Steigerwald and Brus (1989) with PhSe(TMS). The phenyl groups passivate the surfaces and allow the CdSe-capped crystals to be isolated in powder form.

In Fig. 2.12, the syntheses of CdSe nanocrystals in a reverse micelle solution is demonstrated with phenyl groups bonded at the surface. The selenide atoms can be added to the Cd-rich nanocrystal via the PhSe(TMS) or the $Se(TMS)_2$ reagent. The selen is incorporated into the structure of the CdSe nanocrystals but the surface layer is replaced by a permanent chemical 'cap' owing to the strong Se–C bond (Steigerwald and Brus 1989). A similar method has been proposed by Yanagida et al. (1990) for the preparation of CdS nanocrystals. Phenyl-capped CdS particles were obtained during the reaction of Cd^{2+} with $S(TMS)_2$. Then the surface was stabilized in situ with PhS(TMS) under adjustment of the $S(TMS)_2$/PhS(TMS) ratio. The precipitated CdS can either be isolated as stable powder or subsequently redissolved back to many organic solvents. Likewise, Herron et al. (1990) prepared CdS nanocrystals where the final size of the nanocrystals is controlled by adjusting the relative concentrations of thiophenolate and sulfide reagents. It is

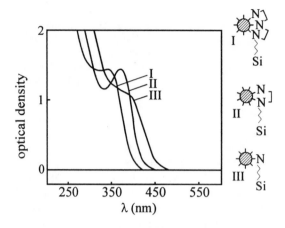

Fig. 2.13. The effect of amine-derived stabilizers (monoamine-, diamine- and triamine-groups) on the optical spectra of CdS cluster sols. From Spanhel et al. (1992a)

possible to use polymer films as a stabilizer for II–VI nanocrystals. Usually, the film containing Cd^{2+} ions has been prepared first. Then S^{2-} ions are added by exposure of the transparent film to H_2S or sodium sulfide. The incorporation of CdS nanocrystals in polymers has been reported both for poly(vinyl alcohol) (PVA) (Woggon et al. 1993a; Nosaka et al. 1993) and for poly(acrylonitrile-co-styrene) (AS) (Misawa et al. 1991). Transparent CdS doped poly(methyl methacrylate) (PMMA) films have been obtained after dissolution of the phenyl-capped CdS nanocrystals in pyridine mixed with PMMA followed by drying onto quartz plates (Yanagida et al. 1990).

A further example of controlled growth of nanocrystals is the impregnation of porous materials with semiconductor inclusions and the use of sol-gel techniques. Suitable porous materials naturally occur as zeolites, a crystalline Al–O–Si material with a network of pores (Wang et al. 1989; Herron et al. 1989; Spanhel et al. 1992a; Reisfeld 1992; Nogami et al. 1990; Kang et al. 1994; Fick et al. 1995). The cavities inside the material are in the order of 1 nm with access through windows of several Ångstroms size. On the one hand, the ordered array of the cavities can be considered a new interesting topic; on the other hand, the onset of mutual interaction of the clusters at a volume fraction of about 20 % makes the understanding of these structures more difficult. It should be mentioned that an attempt has also been made to incorporate III–V semiconductors into porous materials, such as GaAs and InP (Justus et al. 1992; Hendershot et al. 1993; Salata et al. 1994).

Finally, we want to mention the synthesis of CdS nanocrystals by sol-gel based techniques (Bagnall et al. 1990; Spanhel et al. 1991, 1992a,b; Kang et al. 1994; Fick et al. 1995). CdS or ZnO nanocrystals have been stabilized by inorganic multifunctional ligands as demonstrated in Fig. 2.13. These groups act both as link to the semiconductor nanocluster, covalently bonded via short organic chains as well as end groups incorporated into the inorganic network. The nature of the different stabilizing groups strongly influences the electronic states. Drastic changes in the optical spectra have been observed when using different stabilizers derived from the amine family, as shown in Fig. 2.13. The shift of the absorption peaks can be explained by different sizes and probably also by different charge states at the interface.

2.2.2 Size-Selection Techniques

Large efforts are presently being made to prepare highly monodisperse na-nocrystals. As mentioned above, microheterogeneous organic environments already allow a first control of the particle sizes developing (Steigerwald and Brus 1989). In the following, we will discuss two more ways toward the syn-thesis of nearly monodisperse nanocrystals; these are the gel-electrophoresis successfully applied by Weller (1991) and the fast nucleation/slow growth technique recently presented by Murray et al. (1993).

The gel-electrophoretic size separation is a postpreparative size-selective method. Here, nanocrystals are incorporated into a pillar of a suitable gel [e.g. poly(acryl amide)]. Then an external electrical field is applied resulting in a slow migration of the nanocrystals. The velocity of the motion through the pillar depends on the surface charge and size of the nanoparticles. The smallest particles will first arrive at the lower end of the pillar. The gel can be cut into small slices containing different sizes of nanocrystals. Afterwards, the particles can again be dispersed in various solvents. Gel-electrophoretically prepared CdS nanocrystals of 5.5 nm diameter average size show a standard deviation of only 7 % (Weller 1991).

The providing of experimental conditions allowing fast nucleation and slow growth combined with size-selective precipitation has resulted in nearly monodisperse CdSe nanocrystals, as reported by Murray et al. (1993). The reduction of the nucleation process to a short time interval is attained by a rapid increase of the supersaturation upon injection of room-temperature re-agents into a hot coordinating solvent. The sudden temperature drop and the fast decrease in the concentration of the reagents prevent further nucleation. The onset of the slow growth process is promoted by moderate reheating. Si-multaneously to the growth, the absorption spectra are used to monitor the size changes of the crystallites. In response to that, the growth temperature is adjusted to higher or lower values. The production of nanocrystals with larger sizes covering also the weak confinement range has been achieved by use of higher reaction temperatures. To do this, the TMS based precursors for the chalcogen sources have been replaced partly by mixed phosphine/phosphine oxide solutions (TOP, TOPO). The TOP/TOPO is a coordinating solvent with a high boiling point which allows slow steady growth at temperatures above 280 °C. The surface of the nanocrystals is stabilized by the phosphine oxide/chalcogenide moities connected to alkyl groups. The stability strongly depends on the interaction of the alkyl groups with the solvent. When, for example, methanol is added to the solvent, the average polarity increases and the energy barrier for flocculation lowers. The largest particles in the solvent have a higher probability of overcoming this barrier and therefore

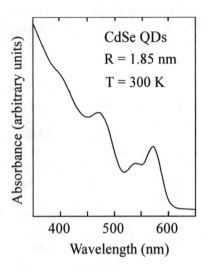

Fig. 2.14. Room temperature linear absorption spectra of R=1.85 nm CdSe nanocrystals with nearly monodisperse sizes. From Murray et al. (1993)

are enriched in the flocculate. The gradual addition of a nonsolvent resulting in a size-dependent flocculation has been exploited to narrow further the particle size-distribution. This subsequent size-selective preparation finally results in highly monodisperse nanocrystals with size deviations < 5 %. Figure 2.14 shows a corresponding absorption spectrum for CdSe nanocrystals. The CdSe samples thus developed already show well-defined structures in the linear absorption. Normally, peaks in the optical spectra can only be detected by means of modulation or size-selective optical techniques (see Chap. 3). The application of this precipitation procedure has been demonstrated for a whole set of II–VI compounds by Murray et al. (1993). Very recently, the self-organization of CdSe nanocrystallites into three-dimensional quantum dot superlattices was successfully achieved and has opened the way to realize ordered arrays of nanocrystals from a variety of II–VI materials (Murray et al. 1995; Kagan et al. 1996).

2.2.3 Sandwiches and Quantum Dot Quantum Wells

A further very interesting development and certainly a topic of intensive future research is the combination of two different semiconductor nanocrystals either by concentrically layered crystallites or by the close contact of two different semiconductor particles connected by fragments at their surfaces. Assuming an abrupt interface, the effect of layering introduces a step function into the Hamiltonian corresponding to the difference in the band-gap energies of the participating semiconductors. In dependence on the order of the materials very interesting changes in the electronic states and in the electron and hole localization can be expected.

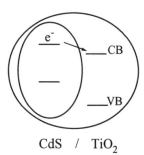

CdS / TiO$_2$

Fig. 2.15. Schematic drawing of sandwich colloids

One example of sandwich colloids has been proposed by Weller et al. (1984, 1991) and Spanhel et al. (1987b), using a combination of CdS and TiO$_2$ (or ZnO). These CdS/TiO$_2$ sandwich structures form after mixing the two components in alkaline solution under high excess of Cd^{2+} ions. In such a structure, the two different semiconductor particles are closely connected, as shown in Fig. 2.15. The light absorbing semiconductor has the smaller band gap, but the absolute energy position of the lowest confined electron level is higher than in the wide gap material. The lowest confined hole state, on the other hand, is situated within the band gap of the attached second semiconductor material. This combination is very similar to the concept of type-II semiconductor structures in two-dimensional confined systems (Bastard 1988). An important result of such a sandwich structure is the spatial charge separation of electrons and holes as a result of electron transfer processes, as observed by Weller et al. (1984) and Spanhel et al. (1987b). Composite semiconductor nanocrystals of CdSe grown on ZnS seeds have been synthesized by Kortan et al. (1990), where the interior core of, for example, CdSe (bandgap of 1.7 eV at 300 K) is surrounded by a closed concentric shell of ZnS (bandgap of 3.2 eV at 300 K) and vice versa. Although CdSe has a cubic lattice constant of 0.608 nm and ZnS one of 0.54 nm, thus implying high strain, fairly uniform growth of the outer layer was demonstrated. An effective thickness of the surface layer of 0.4 nm was achieved. This small layer already causes considerable changes in the optical spectra of the nanocrystals. The successful synthesis of the first quantum dot quantum well structure (QDQW) was reported by Schooss et al. (1994) and Mews et al. (1994). A QDQW is a three-layered structure (schematically shown in Fig. 2.16) which consists of a size-quantized CdS nanocrystal as the core, surrounded by a complete layer of HgS on the surface and covered again by a layer of CdS as the outermost shell. HgS is the semiconductor with the smaller bandgap ($E_{\mathrm{G}} = 0.5$ eV) and is therefore embedded between the core and the outer shell of the material with the larger bandgap, here CdS ($E_{\mathrm{G}} = 2.5$ eV). The dimensions of the layers are only a few nanometers (CdS-core: $R = 2.3$ nm, HgS-layer: $d = 0.3$ nm, CdS-outer shell: $d = 1.4$ nm, see Fig. 2.16) yielding a total diameter of 8 nm for the whole concentric composite particle. The spectrum of the linear ab-

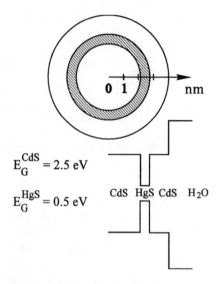

$E_G^{CdS} = 2.5$ eV

$E_G^{HgS} = 0.5$ eV

CdS HgS CdS   H$_2$O

Fig. 2.16. Schematic drawing of CdS/HgS/CdS quantum dot quantum well structures

sorption of the QDQW differs considerably from the sum of the linear spectra of each of the components. The transition energy can be tuned by the core radius, the thickness of the well, and the thickness of the outermost shell. The new lowest electronic transition lies between the bulk excitonic energies of CdS and HgS.

To summarize this section, both preparation methods reported in Sects. 2.1 and 2.2, the growth in glass matrices and the growth in organic or related matrices, have their advantages. To achieve fast nucleation and slow growth of the nanocrystals, the chemical preparation methods using coordinating solvents seem to be very promising. For this type of growth a variety of post-preparative size-selection methods are available. Likewise, in liquid matrix materials the possibility of charge transport exists and charges can be manipulated. Monolayers can be fabricated and put onto conducting surfaces. However, nanocrystals grown in liquid environments are often very unstable and show photochemical reactions. Likewise, these species have to be protected against agglomeration. The necessary surface capping and stabilization can sometimes prevent growth up to sizes considerably larger than the bulk excitonic Bohr radius. Therefore, only a few studies on nanocrystals in organic matrices also covering the weak confinement range have been published.

Glassy matrices are very stable and robust, of good optical quality, compatible with integrated optics and can be considered as preferred materials if the non-uniformity of sizes causes no further disturbance. They can be modified by ion exchange reactions yielding waveguide structures. Manufacture of parallel endfaces and coating with dielectric mirrors is possible. The partial disadvantage of the larger distribution in the average size of the nanocrystals ($\geq 10\%$) can be avoided by applying suitable techniques of nonlinear optics which are able to resolve single nanocrystal properties (see Chaps. 3 and 5).

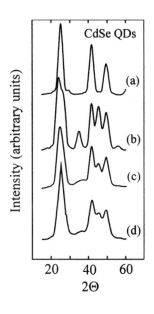

Fig. 2.17. X-ray powder diffraction spectra of 3.5 nm diameter spherical CdSe nanocrystals: (a–c) simulated spectra of (a) pure zinc-blende, (b) pure wurtzite, (c) wurtzite with one stacking fault. Spectrum (d) is the experimental spectrum. From Murray et al. (1993)

2.3 Structural Data

In this section the topical information will be briefly summarized concerning the stoichiometry and the crystal structure of II–VI nanocrystals as well as the data concerning the volume fraction of semiconductor material inside the doped glasses or organics [the parameter p in (1.1)]. As far as available, the data have been collected both for glass and organic matrix materials.

Crystal Structure and Lattice Constants. Almost all analysis based on X-ray measurements gives clear evidence that CdS and CdSe quantum dots predominantly adopt the hexagonal wurtzite-type structure.

For example, the high-resolution TEM and X-ray diffraction applied by Murray et al. (1993) yielded the wurtzite crystal structure with the lattice constant approaching that of the bulk material for all samples of CdS, CdSe, and CdTe nanocrystals. Additionally, larger nanocrystals exhibit a slightly prolate shape with the aspect ratio 1:1.1 to 1:1.3 for the main axes. Stacking faults in the (002) direction have been found. Experimental X-ray powder diffraction spectra for CdSe crystallites have been compared with calculated spectra for 3.5 nm diameter CdSe. The fit of the curves by wurtzite-type crystal structure and adding a single stacking fault gives the best approximation [Fig. 2.17, from Murray et al. (1993)]. Similar results have been reported by Bowen Katari et al. (1994). However, the formation of stacking faults could be avoided by growth at higher reaction temperatures. Besides the absence of defects, the nanocrystals now exhibit spherical shape and again wurtzite-type structure.

Exceptions from the wurtzite-type crystal structure are reported only for the case of extremely small nanocrystals, where a mixture of zinc-blende and

Table 2.3. Comparison of the lattice constants of wurtzite-type CdS and CdSe nanocrystals embedded in glass with radius $R = 4\,nm$ with bulk data, according to Borelli et al. (1987)

	a_{bulk} (Å)	c_{bulk} (Å)	a_{QD} (Å)	c_{QD} (Å)
CdS	4.136	6.713	4.028	6.547
CdSe	4.299	7.010	4.225	6.866

wurtzite-type nanocrystals has been found by Bawendi et al. (1989) and for a number of studies investigating CdS nanocrystals. Weller (1991) found the zinc-blende-type crystallographic structure for 5.5 nm diameter CdS. Likewise Colvin et al. (1992b) observed the zinc-blende structure for CdS between 1 and 10 nm radius.

Under high pressure CdSe nanocrystals can be converted from wurtzite to rock salt structure as observed by Tolbert et al. (1994a,b). This solid–solid phase transition is completely reversible. This very interesting behavior would imply that simultaneously with the phase transition, the transition from a direct-gap to an indirect-gap semiconductor proceeds offering interesting changes in the intrinsic electronic properties of these CdSe nanocrystals (see also Chap. 3).

In glasses, only the bulk-like wurtzite-structure has been observed, in particular for radii larger than 2 nm (Liu and Risbud 1990; Potter and Simmons 1988; Borelli et al. 1987; Champagnon et al. 1993). By high-resolution TEM the evolution of the shape of the CdSe nanocrystals embedded in glass was studied by Champagnon et al. (1993). In the early stage of growth ($R \leq 6\,nm$) the nanocrystals are nearly spherical. Larger nanocrystals ($R > 8\,nm$) grow more and more in the direction of the c-axis, become prolate and the ratio between the two main axes amounts to ~20 %. Later, the crystals come close to hexagonal prisms.

Lattice constants for CdS and CdSe nanocrystals measured by Borelli et al. (1987) are given in Table 2.3. The lattice constants are slightly smaller than those of the bulk material, indicating some compressive strain. Compressive or dilative strain is a type of lattice relaxation decreasing the number of dangling bonds and retaining the original crystal lattice. Compression appears if the outer bonds at the nanocrystal surface are shorter than the inner ones. The existence of compressive strain for nanocrystals in glass has been confirmed by Raman scattering and explained by the size dependence of the surface free energy by Scamarcio et al. (1992). To a smaller extent the differences in the thermal expansion coefficients between the glass and the semiconductor material contribute to the strain. For CdS_xSe_{1-x} with $x = 0.65$ and $R = 4.3$ nm a strain of $\Delta a/a = -9.4 \times 10^{-3}$ has been determined.

A further possibility to identify the wurtzite structure is the measurement of the luminescence polarization. In a hexagonal crystal, the degeneracy of the valence band at $k=0$ is lifted. The transition from the A-valence band to the conduction band is allowed for light-polarization vectors perpendicular to the c-axis whereas the transition from the B-valence band is allowed for both polarizations, but with different probabilities. The luminescence spectrum was measured by Chamarro et al. (1992) for $CdS_{0.4}Se_{0.6}$ nanocrystals embedded in glass with an average length of 37 nm in the direction of the c-axis and a width of 26 nm. Partially polarized light was detected. The application of luminescence polarization measurements for characterizing the lattice type is certainly a good tool for nanocrystals of large sizes. As shown later in Sect. 3, for small-size nanocrystals the discussion of polarized luminescence is more sophisticated (Efros 1992a).

Volume Fraction. For the CdSe nanocrystals grown at $580\,°C/80$ h and $600\,°C/170$ h (see Müller et al. 1992) the filling factor $p = V_{CdSe}/V_{total}$ has been calculated from the scattering intensity of SAXS experiments. For the sample treated at $580°C$, a value of $p = 2.9 \times 10^{-3}$ was found and for the sample treated at $600\,°C$, the value is $p = 3.8 \times 10^{-3}$. The lower value of p for the $580\,°C/80$ h sample confirms the assumption that the growth within this low-temperature regime occurs via precipitation of dissolved ions from the matrix. Similar values for p around $0.5\,\%$ up to a maximum of $1\,\%$ have also been reported by other authors (Potter and Simmons 1988; Liu and Risbud 1990). To examine the size distribution and volume fraction in CdS_xSe_{1-x} mixed crystals Goerigk et al. (1994) applied a contrast variation technique obtaining values between $p = 0.007$ ($CdS_{0.4}Se_{0.6}$, $R = 5 \pm 1.8$ nm) and $p = 0.03$ ($CdS_{0.6}Se_{0.4}$, $R = 4.6 \pm 1.7$ nm). These values correspond to a mean crystallite number density of 1.3×10^{16} cm^{-3} and 7.5×10^{16} cm^{-3}, respectively. Somewhat larger values ($p = 1$–$10\,\%$) can be attained by embedding the nanocrystals into polymers. For the growth of CdS in PVA a volume fraction around $3\,\%$ has been reported (Woggon et al. 1993a).

Stoichiometry. To study the intrinsic physical properties of nanocrystals, there is a need for samples with well-defined stoichiometry. In the diffusion-controlled growth process the velocity of transport through the matrix is different for the ions taking part, due to the different radii. From this point of view, it is very likely that the mixed crystal composition x, e.g. for S and Se, is different when using different growth temperatures, even for identical matrix material. The same problem occurs for the Zn–Cd ratio in Zn-based glasses (Yükselici et al. 1995). Hence, nanocrystals with widely varying stoichiometry occur in commercial filter glasses. This effect can be partly compensated and the change in stoichiometry minimized by adding ions of larger radii, e.g. Sn ions, and a suitable process control through the growth temperature (Yan and Parker 1989). For the organic matrix environment, however, it seems to be very easy to adjust equal stoichiometric ratios between Cd and Se as shown by Bowen Katari et al. (1994).

2.4 Influence of Interfaces

The wish to confine the motion of electrons and holes in the nanocrystal very strongly is always connected with small sizes of the nanocrystals and by this the increasing influence of interfaces. Therefore, it is worthwhile to refer to the role of interface properties here and in the sections below.

The question of surface formation, of the kind of atoms at the outer or inner part of the interface, is still the subject of intensive investigations. Likewise, the extension of the surface zone disturbed by dangling bonds, vacancies or impurities, is not completely solved as well as the mechanism of charge compensation of ions at the nanocrystal surface. Growth regimes determined by a diffusion-controlled process are frequently characterized by a gradient in concentration around the cluster surface. The competitive growth between the subcritical and supercritical clusters results in a depletion or an excess of ions in the direct surrounding of the nanocrystals. The best condition is the case of $R = R_{cr}$, the equilibrium state. Analyzing the growth process, the improvement of the interface quality is an essential part of present research. The surface morphology not only plays an important role concerning the stability of the nanocrystals, it also influences the emission energies and the attainable quantum yield. The species capping the surfaces are also thought to passivate possible surface trap states. To establish the structure of the surface shell, nuclear magnetic resonance (NMR), X-ray photoelectron spectroscopy (XPS) or luminescence experiments have been applied successfully.

The surface morphology of CdSe nanocrystals has been investigated by Becerra et al. (1994) exploring the ^{31}P nuclear magnetic resonance. As expected from the preparation method after Murray et al. (1993), two surface-capping species were identified, the TOPO and TOPSe groups. The average crystallite surface was surrounded by a nearly closed packed hydrophobic shell with 70 % TOPO molecules and 30 % TOPSe molecules passivating Cd sites. It was suggested that uncoordinated Se sites act as potential hole traps.

X-ray photoelectron spectroscopy was applied by Bowen Katari et al. (1994) to characterize the surface of CdSe nanocrystals, likewise synthesized after Murray et al. (1993), however bonded to gold or silicon conducting surfaces for study (Colvin et al. 1992b). The results are in good agreement with the NMR studies presented by Becerra et al. (1994). The majority of Cd surface atoms is bonded to the TOPO ligand whereas the Se surface sites remain unbonded. The observed degradation over time is attributed to oxydation of Se to SeO_2. Moreover, if pyridine is used for deposition neither the Cd nor the Se sites can be completely passivated and oxydation occurs for both species upon exposure to air.

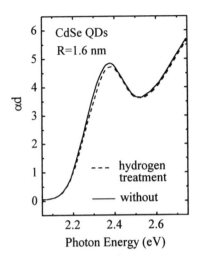

Fig. 2.18. Influence of hydrogenation of the interface of small CdSe nanocrystals embedded in glass. From Woggon et al. (1990)

The formation of a layer of $Cd(OH)_2$ was observed by Spanhel et al. (1987). Cd^{2+} ions are bound to sulfide anions and passivate the surface imperfections thought to be responsible for trapping. As a result, after surface modification a bright green-blue luminescence appears with a quantum yield exceeding 50 %.

The luminescence of CdS nanocrystals has been analyzed by Misawa et al. (1991) with respect to the role of the Cd salts involved during the preparation process. A counter-ion-dependence has been established by comparing luminescence spectra of various samples. According to these observations a site-substitution model has been proposed where the trapped states yielding red luminescence are impurities substituting the sulfur sites.

In glassy matrices the chemical control of interfaces is more difficult to realize. According to Borelli et al. (1987) small CdSe nanocrystals embedded in borosilicate glasses were characterized by a Cd-rich surface. To compensate charge states, defects or dangling bonds, hydrogenation is widely used and known from silicon technology to passivate Si–SiO_2 interfaces. Woggon et al. (1990) studied the influence of the surface passivation by exposing CdSe-doped glasses to a hydrogen atmosphere at 300 °C for 4 hours. After the hydrogenation, which removes surface related dangling bonds and non-bridging O-bonds, the absorption peak in the linear spectrum is slightly shifted to higher energy (see Fig. 2.18). Even though the changes in the linear absorption are only marginal, the change in the nonlinear absorption and the photodarkening is remarkable, as will be discussed later in Chap. 6.

It can be concluded from this discussion that two groups of nanocrystals are advantageous for optical experiments, namely small nanocrystals with stabilized surfaces or very large ones with a corresponding small surface-to-volume ratio and, hence, minor importance of interface properties.

2.5 Epitaxial Growth

A completely different approach to obtain quasi-zero-dimensional semiconductor structures starts from epitaxial growth techniques. Two principal ways can be distinguished, (i) the growth of a two-dimensional layered structure (quantum well) followed by successive etching procedures, thus further lowering the dimension, and (ii) epitaxial techniques exploiting self-organization and island-like growth within the first deposited monolayers.[1]

For structuring semiconductor materials by etching, ion or electron beam techniques are widely used. Advanced technologies reach structures in the sub-100 nm range. After covering the quantum well structure with electron sensitive positive or negative resists, the surface is exposed to the electron beam. During the exposure the resist changes its chemical structure and thus its solubility. In the following dry-etching procedure the material, which was not protected by the resist, will be removed. The dimension of the remaining material is determined by the exactness of inscribing the mask pattern to the resist (proximity effect and scattering), as well as by etching below the resist and thereby damaging the side walls of the structure. In the case of free-standing columns, the vacuum is the confining potential for the electrons and holes, in the case of a subsequent overgrowth of the etched structures the quantum dots are buried and the surrounding material presents the potential barrier. For both cases very different electronic properties are expected, as outlined in more detail in Chap. 8.

Originally, etching procedures were installed to realize micron size mesa structures, but over time they have been driven down to the range of $\sim$25 nm (for more details see, for example, Forchel et al. 1988; Kash 1990; Reed and Kirk 1991; Sotomayor Torres et al. 1992; Heitmann and Kotthaus 1993). After such a structurization, III–V semiconductors (with typical values of the excitonic Bohr radius around $\sim$100 nm) should become small enough to exhibit remarkable quantum confinement. Some examples for realized dry-etched structures are the 60 nm diameter $GaAs/Al_xGa_{1-x}As$ dots reported by Wang et al. (1992a) and the 40 nm diameter $In_{0.17}Ga_{0.83}As/GaAs$ dots obtained by Daiminger et al. (1994).

For II–VI semiconductors a different situation exists. Here the excitonic Bohr radius is smaller and in the range of a few nanometers only. Although the lateral dimensions of the etched structures reach the range of a few tens of nanometers, the confinement effect remains weak. Often the confinement-induced small high-energy shifts of the resonances are covered by other effects, like strain-induced energy shifts. Nevertheless, first successful etching of II–VI dot- and wire-like structures with high optical quality were published for

---

[1] The additional application of electrostatic or magnetic fields, similarly, can produce confinement potentials and is a further way to obtain quantum dots, in particular in III–V based nanostructures. This topic will be the subject of a later chapter.

CdZnSe/ZnSe heterostructures by Illing et al. (1995), and for CdTe/CdZnTe heterostructures by Gourgon et al. (1995).

A further idea to derive quantum dots from quantum well structures has been proposed by Brunner et al. (1992). By a focused Ar^+ laser beam (temperature of about $1000\,°C$, spot $500\,nm$) a local Al/Ga interdiffusion has been promoted which locally increases the confining barrier and yields so-called buried quantum dots of dimensions between $250\,nm$ to $2000\,nm$.

Over the years the semiconductor system InAs grown on GaAs(100) has become a well-established model system for the study of the mechanism of heteroepitaxy in the case of highly lattice-mismatched semiconductors (see e.g. Goetz et al. 1983; Grundmann et al. 1990 and references therein). For InAs/GaAs(100) the lattice mismatch is $\sim 7\,\%$. The formation of dislocation-free coherent islands of InAs on the GaAs substrate has been observed and attributed to a transition from a two-dimensional growth mode to a three-dimensional mode [model by Stranski and Krastanow (1939)]. The three-dimensional growth mode (either as the change from the layer growth to the island growth or as the direct nucleation in islands at the beginning of the deposition) is now again at the center of interest and recovered as a technique of so-called self-organized growth to obtain regularly shaped quantum dots. Whether the two-dimensional or three-dimensional growth mode dominates the growth process depends very sensitively on the magnitude of the lattice mismatch (strain) between the substrate and layer material, the crystallographic orientation of the substrate, the substrate roughness and temperature, the elastic constants of the materials involved and the surface energy of the nanocrystal geometries grown. During the nucleation stage, atoms impinge upon the substrate, diffuse across the surface, form subcritical clusters with other atoms which gradually coalesce and develop towards a two-dimensional wetting layer. On top of this wetting layer coherently strained islands grow until they reach a critical thickness. The dispersion in size and shape of these islands is very small at this stage of growth. Above the critical thickness, the lattice mismatch will be relieved by the generation of dislocations at the interface between layer and substrate giving an effective strain relaxation. The incoherent, strain-relaxed islands, once formed, continue to grow without any restriction until the complete two-dimensional film has been developed.

First proposals for exploiting the lattice-mismatch in the $In_xGa_{1-x}As/$ GaAs system in order to realize three-dimensional confinement have been made, for example, by Leonard et al. (1993). The results reported in the literature show surprisingly uniform behavior concerning the base length and shape of the grown islands or the value of critical coverage of monolayers for the onset of island growth. Some examples will be listed in the following. Moison et al. (1994) report nearly pyramidal quantum dots of $3\,nm$ height and $12\,nm$ half-base with a dispersion of sizes of only $\pm 15\,\%$. The critical coverage, above which the three-dimensional growth starts, is 1.7 monolayers and the average interdot distance $60\,nm$. Similar results have been obtained by Marzin et al. (1994a). Here InAs islands are nucleated on the top surface of a 2D InAs layer after reaching the critical thickness of 1.7 monolayers. Then the growth is interrupted to wait for the evolution of these islands into equilibrium. The results are InAs islands of 2.8 nm height, 24 nm diameter and 15 % size fluctuation. For the inter-island distance, a value of 55 nm has been obtained. Leonard et al. (1994) showed that the most uniform InAs islands ($d \approx 25$ nm) can only be produced at the very initial stages of their formation at approximately 1.6 monolayers InAs coverage. The nucleation rate was found to depend critically on the morphology of the substrate surface, i.e. on the density of monolayer-height steps of the GaAs substrate.

When mixed crystals are grown the parameter of critical coverage changes. As demonstrated by Ruvimov et al. (1995) and Grundmann et al. (1995a), the critical thickness for the formation of three-dimensional islands changes with x from 1.7 monolayers to 3 monolayers for $x = 1$ to $x = 0.5$. Furthermore, differences in the dot density and shape occur with increasing x (spheres for $x = 0.5$, pyramids with square base for $x = 1$). In the case of $In_{0.5}Ga_{0.5}As$, small pyramids develop with 12 nm base length and 6 nm height. They show a short range ordering in chains and the formation of a two-dimensional cubic lattice. The reported size dispersion of dots is $< 20\%$. A value of 10^{11} dots/cm^2 has been given for the dot density.

The MBE growth of $In_xGa_{1-x}As/GaAs(100)$ under variation of the x content has also been analyzed by Gerthsen et al. (1994), and Tillmann et al. (1995). Here the early stages of growth have been investigated and the relaxation of lattice mismatch by formation of defects have been studied. For the layer thickness leading to an effective strain relaxation, a value of ≥ 3.5 nm has been determined, corresponding to island sizes of around 50 nm. Thus the first misfit dislocation occurs approximately after three monolayers. Further investigation of the formation process of misfit dislocations can be found, for example, in the work of Grundmann et al. (1990), Chen et al. (1996), and Zou et al. (1996) and references therein.

For the composition $x = 0.55$ of the ternary compound $Al_{0.45}In_{0.55}As$ grown on GaAs, the formation of quantum dots with 17 nm base dimensions, 3 nm height, a size dispersion of $\pm 12\%$, and a dot density of 200 μm^{-1} has been observed by Fafard et al. (1994a).

Since both size and shape variation as well as changes in the mixed crystal composition x lead to shifts in the optical absorption features, the exact knowledge of the InAs/GaAs ratio becomes essential for the understanding of the electronic states in the grown structures. Currently, the role of the wetting and cap layer is the subject of intensive investigations and the migration of atoms out of these layers into the dots is being studied. It has been shown that the size and volume change of the islands involves In atoms from the two-dimensional wetting layer and that the dot formation results in a decrease of the In content of the wetting layer (Leonard et al. 1994; Tillmann et al. 1995; Krost et al. 1996).

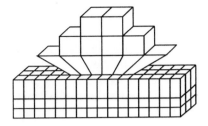

Fig. 2.19. Schematic illustration of the three-dimensional growth mode during the epitaxial deposition of semiconductor material with strong lattice mismatch with respect to the substrate

We are now turning to the theoretical description of the dot growth process as schematically illustrated in Fig. 2.19. The first theoretical treatment of the epitaxial growth applied the homogeneous nucleation theory to the growth process of layers. In analogy to the Lifshiz–Slesov model the nucleation process on layers and its temporal development has been modeled by Chakraverty (1967). However, the experimentally observed size distributions (see the examples above) are generally much smaller than the ones predicted from an extension of the Lifshiz–Slesov model to epitaxial growth techniques. Therefore, Priester and Lannoo (1995) proposed a model that includes the interplay of surface energy and elastic energies and explains the existence of a critical value for the coverage above which the three-dimensional island-growth process dominates. In their model, at first two-dimensional platelets are formed on the substrate. Afterwards, three-dimensional [104] pyramidal InAs islands form in dependence on the input values of the relaxation and surface energies. According to their calculations, the following scenario proceeds: The first layer which grows is a wetting layer completely coherently strained with respect to the substrate. Then the atoms deposited on the next layer form two-dimensional platelets with sizes determined by the diffusion length. These platelets have a precursor action and, with increasing coverage above a critical value, the platelets start to interact and the formation of three-dimensional island is energetically more favorable [the value of critical coverage has been given with 1.8 monolayers for the case of InAs on GaAs(100)]. The atoms are more likely to jump to the top of the platelet and form the next monolayer than to sit in the first monolayer between two 'repulsive' platelets. The size distribution does *not* change in this process and is further a narrow Gaussian. The islands subsequently grow to sizes which correspond to the critical monolayer thickness of strain relaxation due to formation of misfit dislocations. At this stage the islands do not continue to develop identically and start to coalesce in a way similar to the one described by the Lifshiz–Slesov model for bulk material. Strain relaxation and transition towards two-dimensional growth occurs and thick layers develop. The smallest island size following this calculation is ∼16 000 atoms with 33 % width of the Gaussian distribution.

Besides the InAs/GaAs(100) combination with the strong lattice mismatch, the formation of quantum dot like structures has been reported for InGaAs grown on tetrahedral-shaped recess patterned on (111)B GaAs substrates (Sugiyama et al. 1995), and for the growth on (311)B substrates where a spontaneous reorganization between AlGaAs and strained InGaAs films has been observed. The results are InGaAs islands buried beneath AlGaAs with sizes from 200 nm down to 30 nm (Nötzel et al. 1994a,b). For GaSb grown on GaAs(100) substrates, regularly shaped quantum dots are formed with ~20 nm sizes for which a type-II confinement structure has been suggested, i.e. with electrons and holes confined separately in different materials (Hatami et al. 1995).

Self-alignment of the quantum dots has been achieved by using specially prepared substrates like etched GaAs ridges (Mui et al. 1995), multiatomic steps on a vicinal GaAs substrate (Kitamura et al. 1995), or patterned (311)A GaAs substrates (Nötzel et al. 1996). The spontaneous alignment of the InGaAs quantum dots (and by this the spatial control of the position of the grown islands) is desirable in certain device structures, for example, for modulating the gain in laser structures.

The different features of luminescence reflecting the electronic properties and pecularities of these types of quantum dots preferentially found in III–V materials will be outlined in more detail in Chap. 8.

3. Energy States

In this chapter we discuss conditions for which a nanocrystal can be considered as a quantum dot. To explain the three-dimensional quantum confinement, the quantum mechanical problem of the motion of a particle in a box will be briefly recalled. This simple model has been improved during the last few years by adding more and more complicated expressions to the Hamiltonian, taking into account, for example, electron–hole Coulomb interaction, a more complicated valence-band structure or nonparabolic bands. Following this historical method, the development of the various models will be briefly sketched and the results of recent theories will be compared with experiments. I do not attempt to review the entire area of theoretical calculations (for this see Banyai and Koch 1993), but we will see that models going beyond simple quantum mechanics give rise to drastic alterations in a number of properties such as selection rules, the radial probability of the electron and hole wavefunctions and, of course, the energy states themselves.

In this and the next sections it will be shown that linear absorption alone is not sufficient to identify unambiguously the energy states of the quantum dots. Only the combination of linear and nonlinear optical methods allows a direct comparison of theory and experiment.

Starting with one-electron–hole-pair (1EHP) states and going to two-pair (2EHP) and many-particle states, we gradually increase the density and population of excited states in the dot. This can be realized, for example, by an increase of the excitation density or by changing the size of the nanocrystals. Therefore, inevitably we have to discuss the strong, intermediate and weak confinement regime although we will not use these terms to organize this chapter.

3.1 One-Electron–Hole-Pair States

3.1.1 The Particle-in-the-Box Model

In the following we consider nanocrystals with a spatial extension larger than the lattice constants. It is very likely that in this range of sizes the crystalline structure of the bulk material has already been developed and the effect of

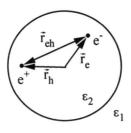

Fig. 3.1. Electron and hole in a semiconductor sphere with the dielectric constant ϵ_2 embedded in a host material with ϵ_1

quantum confinement can be observed. The description of the quantum confinement within the framework of the effective mass approximation (EMA) is a good approach. An alternative way is the treatment within the LCAO (Linear Combination of Atomic Orbitals) approximation. The energy states calculated within a tight-binding approximation coincide sufficiently well with results obtained by EMA for the range of dot radii down to 1.5 nm, as shown by Lippens and Lannoo (1989, 1990), Einevoll (1992), and Ramaniah (1993). Recently, Wang and Zunger (1996) reported results which are going beyond the EMA for the range of small dot sizes. Very good agreement between the results of a pseudopotential calculation and experimental data was achieved for CdSe quantum dots with sizes starting from small clusters ($R = 0.6$ nm) up to CdSe dots of about 1000 atoms ($R = 1.9$ nm).

In bulk crystalline solids the electron states are usually obtained by solving the stationary Schrödinger equation of an electron in a spatially periodic potential

$$\hat{\mathbf{H}} \Psi(\mathbf{r}) = \left[-\frac{\hbar^2}{2m} \nabla^2 + V(\mathbf{r}) \right] \Psi(\mathbf{r}) = E \Psi(\mathbf{r}) \,. \tag{3.1}$$

The potential $V(\mathbf{r})$ has the periodicity of the underlying Bravais lattice

$$V(\mathbf{r}) = V(\mathbf{r} + \mathbf{R}) \tag{3.2}$$

for all lattice vectors $\mathbf{R}$. The eigenstates Ψ of (3.1) were composed from an envelope function, the plane wave e^{ikr}, and a function $u_{\nu,k}(\mathbf{r})$ periodically with the lattice vector of the Bravais lattice

$$\Psi_{\nu,k}(\mathbf{r}) = e^{ikr} u_{\nu,k}(\mathbf{r}) \tag{3.3}$$

and

$$u_{\nu,k}(\mathbf{r}) = u_{\nu,k}(\mathbf{r} + \mathbf{R}) \,. \tag{3.4}$$

Here ν is the band index and $\mathbf{k}$ the wave vector reduced to the first Brillouin zone. Equations (3.3) and (3.4) represent the well-known Bloch theorem. The energy eigenvalues in parabolic band approximation are then given by

$$E(\mathbf{k}) = \frac{\hbar^2 \mathbf{k}^2}{2m} \tag{3.5}$$

with m being the effective mass of the electron or the hole.

We turn now to the problem of the electron–hole-pair states inside a spherical potential (Fig. 3.1). In the most simple treatment, a semiconductor sphere with radius R is surrounded by an infinitely high potential barrier represented by the matrix material. In the framework of the envelope function approximation, it is assumed that the wave function can be expanded in products of the cell periodic parts $u_{\nu,k}$ of the Bloch functions together with a new, specific envelope function. The periodic parts of the Bloch functions are supposed to be the same in the barrier and the well,

$$u_{\nu,k}(r)_{\text{barrier}} = u_{\nu,k}(r)_{\text{well}} = u_{\nu,k}(r) . \tag{3.6}$$

The aim is to determine for the particle-in-a-spherical-box problem the new envelope function ψ for electron and holes,

$$\Psi(r) = \psi(r)\, u(r) . \tag{3.7}$$

The Hamilton operator for the envelope function $\psi(r)$ in single parabolic band approximation, neglecting first Coulomb interaction, is given by

$$\hat{\mathbf{H}} = -\frac{\hbar^2}{2m_e}\nabla_e^2 - \frac{\hbar^2}{2m_h}\nabla_h^2 + V_e(r_e) + V_h(r_h) \tag{3.8}$$

with the confinement potential

$$V_i(r_i) = \begin{cases} 0 & \text{for } r_i < R \quad (i = \text{e}, \text{h}) \\ \infty & \text{for } r_i > R \quad (i = \text{e}, \text{h}) \end{cases} . \tag{3.9}$$

For non-interacting electron–hole pairs, the envelope wavefunction ψ is separable in independent contributions from electrons and holes and can be written by the product

$$\psi(r_e, r_h) = \phi_e(r_e) \cdot \phi_h(r_h) . \tag{3.10}$$

The solution of the Schrödinger equation with (3.8–3.10) can be found in textbooks of quantum mechanics (for example Davydov 1987). The normalized wavefunctions ϕ_i for electrons as well as holes are

$$\phi_{nlm}^i(r) = Y_{lm} \sqrt{\frac{2}{R^3}} \, \frac{J_l\left(\chi_{nl} \frac{r}{R}\right)}{J_{l+1}(\chi_{nl})} \tag{3.11}$$

with $-l \leq m \leq l$; $l = 0, 1, 2, ...$; $n = 1, 2, 3, ...$ Here J_l are the Bessel functions, and Y_{lm} are the spherical harmonics. The energy eigenvalues E_{nl} follow from the requirement that the wave function has to vanish at $r = R$, the boundary quantum dot–matrix,

$$J_l\left(\chi_{nl} \frac{r}{R}\right)\Big|_{R=r} = 0 \tag{3.12}$$

and are determined from this condition as

$$E_{nl}^{\text{e,h}} = \frac{\hbar^2}{2m_{\text{e,h}}} \frac{\chi_{nl}^2}{R^2} . \tag{3.13}$$

χ_{nl} is the n-th zero of the spherical Bessel function of the order l and $m_{e,h}$ are the effective masses of the electron and the hole. R is the radius of the nanocrystals. Labelling the quantum numbers $l = 0, 1, 2...$ with the letters $s, p, d...$, the first roots are $\chi_{1s} = \pi$; $\chi_{1p} = 4.493$; $\chi_{1d} = 5.763$; $\chi_{2s} = 2\pi$; $\chi_{2p} = 7.725$ etc. Using (3.13) it can be easily shown that the quantization energy of the lowest energy state with $n = 1$ and $l = 0$ is given by

$$E_{10}^i = \frac{\hbar^2}{2m_i} \frac{\pi^2}{R^2} . \tag{3.14}$$

The energy of a particle in a spherical potential well takes discrete values and scales with the square of the inverse radius R. With respect to the bandgap of the bulk semiconductor, an increase ΔE of the transition energy

$$\Delta E = \frac{\hbar^2}{2\mu} \frac{\pi^2}{R^2} \tag{3.15}$$

has been obtained for the lowest confined 1EHP state, where μ is the reduced effective mass of the electron–hole pair ($\mu = m_e m_h / (m_e + m_h)$).

The above consideration only deals with the envelope problem. The Bloch parts of the wave functions have been approximated by a simple two-band semiconductor with parabolic, isotropic bands and direct energy gaps with the maximum of the conduction and valence band situated in k-space at $k = 0$. In reality the quantum dot structures are based on a great variety of semiconductor materials having both indirect band gaps, zinc-blende or wurtzite-type crystal structure or are grown in directions with highly anisotropic lattice symmetry. Strictly speaking, the properties introduced by the Bloch part of the wave function $u_{v,k}(r)$ cannot be neglected if one wishes to have a complete understanding of the optical behavior of the quantum dots. We refer to the following sections for this problem.

Optical Transitions. To discuss the optical absorption spectrum $\alpha(\omega)$ resulting from the solution of the particle-in-the-box problem, the probability of dipole-allowed optical transitions between single electron and hole states has to be evaluated,

$$\alpha(\omega) \sim |\langle \Psi_f | \mathbf{e} \cdot \hat{\mathbf{p}} | \Psi_i \rangle|^2 \tag{3.16}$$

with $\mathbf{e} \cdot \hat{\mathbf{p}}$ the dipole operator ($\mathbf{e}$ denotes the polarization). Ψ_i and Ψ_f are the initial and the final states of the optical transition. With the abbreviations,

$$\begin{aligned} |i\rangle : \quad & \Psi_i(r) = u_{vi} \cdot \phi_i(r), \\ |f\rangle : \quad & \Psi_f(r) = u_{vf} \cdot \phi_f(r) \end{aligned}$$

and considering only the 'interband' part ($\nu_i \neq \nu_f$), the overlap integral reads as

$$\langle f | \mathbf{e} \cdot \hat{\mathbf{p}} | i \rangle = \underbrace{\langle u_{vf} | \mathbf{e} \cdot \hat{\mathbf{p}} | u_{vi} \rangle}_{p_{cv}} \langle \phi_f | \phi_i \rangle . \tag{3.17}$$

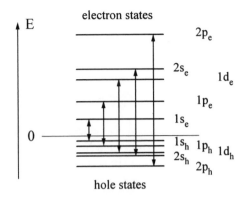

Fig. 3.2. Scheme of the dipole-allowed optical transitions within the non-interacting particle approximation

The integral can be separated into the integration of the fast oscillating Bloch part and the integration of the envelope part. The integration of the Bloch part results in the size-independent interband dipole matrix element p_{cv} of the bulk. The quantum-dot specific selection rules of 'interband' transitions between confined electron and hole states were obtained by integrating the ϕ functions over the quantum-dot volume. These integrals can be computed exactly, making use of the symmetry of the functions ϕ^i_{nlm}. Due to the ortho-normality of the envelope functions ϕ_i and ϕ_f, the integration yields delta-functions δ_{if}. As the result we obtain the well-known selection rule that all transitions that conserve n and l are allowed between the non-interacting electron and hole states, i.e. $1s_e \to 1s_h$, $1p_e \to 1p_h$, $1d_e \to 1d_h$ etc. The oscillator strength of these transitions is proportional to $(2l+1)$ due to the summation over all states with $-l \le m \le l$ contributing to the absorption. Neglecting line broadening effects, we will find optical transitions at the energies $\hbar\omega = E^e_{nl} + E^h_{nl}$ above the band gap and for the absorption coefficient hold

$$\alpha(\omega) \sim |p_{cv}|^2 \frac{1}{\frac{4\pi}{3}R^3} \sum_{n,l} (2l+1)\delta(\hbar\omega - E^e_{nl} - E^h_{nl}) . \tag{3.18}$$

The possible optical transitions are schematically sketched in Fig. 3.2.

Carrying out the same procedure for the 'intraband' transitions, i.e. $\nu_i = \nu_f$, one can verify that the selection rules are $n_f \ne n_i$; $l_f - l_i = 0, \pm 1$; and $m_f - m_i = 0, \pm 1$ for optical transitions within the ladder of either electron or hole states.

This simple consideration is good for an initial and, more qualitative discussion of, for example, the blue shift of the lowest transition in the absorption of small nanocrystals [or now quantum dots (QD)], but it cannot explain in detail the observed change in the absorption spectra resulting from the quantum confinement.

3.1.2 Coulomb Interaction

The next step towards an exact treatment is the inclusion of the Coulomb interaction between the electron and the hole inside the quantum dot. This problem corresponds to the following Hamiltonian

$$\hat{H} = -\frac{\hbar^2 \nabla_e^2}{2m_e} - \frac{\hbar^2 \nabla_h^2}{2m_h} - \frac{e^2}{\epsilon_2 |r_e - r_h|} + V_e(r_e) + V_h(r_h) . \tag{3.19}$$

In bulk materials the influence of the Coulomb potential is essential and is the reason for the existence of the excitons. The bulk Hamiltonian however [(3.19) without the confining potentials V_e and V_h], is separable into the relative and center-of-mass motion of the electron–hole pair. In spherical quantum dots, taking into account the Coulomb potential, there is a break in symmetry, because the Coulomb interaction depends on the spatial distance between the electron and the hole. Since the separation of (3.19) in coordinates of the relative and center-of-mass motion is no longer possible in a simple way, analytical solutions are difficult. Therefore, the most simple approach takes into account that the Coulomb energy scales like the inverse of the electron–hole distance ($\sim 1/R$), whereas the kinetic energy scales like the square of the inverse radius ($\sim 1/R^2$). One possible description for small dot radii in the so-called strong confinement range ($R \ll a_B$, where a_B is the excitonic Bohr radius) is the neglect of all Coulomb interaction effects, and the solution of the problem reduces to the treatment discussed above within the frame of noninteracting electrons and holes. Similarily, for quantum dot sizes $R \gg a_B$ the problem can again be solved in analogy to the particle-in-a-box problem if one considers the electron–hole pair, i.e. the exciton, as single particle with the confining potential acting only on the center-of-mass motion. This so-called weak confinement case gives a good approximation in semiconductors with high exciton binding energies such as CuCl and CuBr which have excitonic Bohr radii of 0.7 nm and 1.25 nm, respectively.

Despite these crude approximations, the Hamiltonian (3.19) has been treated both by perturbation theory (Brus 1984), variational calculations (Schmidt and Weller 1986; Ekimov et al. 1989; Nair et al. 1987; Kayanuma 1988, Kayanuma and Momiji 1990; Tran Thoai et al. 1990; and Takagahara 1993a), Monte-Carlo-technique (Pollock and Koch 1991) and matrix diagonalization methods (Hu et al. 1990b, Park et al. 1990). Treating the problem in perturbation theory (Brus 1984), the lowest excited state energy is given by

$$E_{10} = \frac{\hbar^2 \pi^2}{2R^2} \left[\frac{1}{m_e} + \frac{1}{m_h} \right] - \frac{1.8 \, e^2}{\epsilon_2 \, R} . \tag{3.20}$$

Going from the non-interacting particle picture to the electron–hole-pair picture, the description of the corresponding optical transitions changes from Fig. 3.2 to Fig. 3.3. The inclusion of the Coulomb potential diminishes the transition energy between the $(1s_e, 1s_h)$-pair state and the crystal ground state.

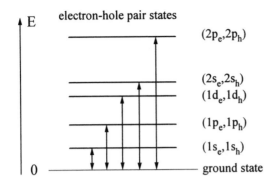

Fig. 3.3. Scheme of the optical transitions in the frame of the one-electron–hole pair (1EHP) picture

To avoid the problem of a break in symmetry for the intermediate range of confinement ($a_B \approx R$), Ekimov et al. (1989) have proposed an ansatz where the hole is located at the center of the quantum dot. From this model of the donor-like exciton inside a quantum dot, a variational calculation has been started and the ground state energy has been determined as a function of size. The localization of the hole has been assumed to be the result of the potential of the electron moving in the confining quantum dot potential.

When using the numerical method of variational calculations, the creative part of the work consists in finding a bright idea for the trial functions. Variational calculations have been performed for single parabolic band structures and infinitely high barriers by Nair et al. (1987) and Kayanuma (1988). The trial wave functions introduced are, on the one hand, products of the solutions for non-interacting particles involving lowest-order Bessel functions or, on the other hand, an exponential function taking its pattern from the 1s-wave function of the hydrogen problem. Up to three parameters have been introduced and the variational function has then been optimized with regard to these parameters. Nair et al. (1987) calculated the energy of the lowest electron–hole-pair state variationally using three-parameter Hylleraas–type wave functions. Kayanuma (1988) used a polynomial basis for the trial wave function and the expansion coefficients as the variational parameters. For the limit of strong confinement an analytical formula useful for a first-principle estimation has been given in this work. Assuming the Coulomb potential as a small perturbation, the author derived an expression similar to (3.20)

$$E_{10} = \frac{\hbar^2 \pi^2}{2R^2} \left[\frac{1}{m_e} + \frac{1}{m_h} \right] - \frac{1.786\,e^2}{\epsilon_2\,R} - 0.248 E_{\mathrm{Ryd}}^* \tag{3.21}$$

with E_{Ryd}^* the bulk exciton binding energy in meV.

However, both in the work of Nair et al. (1987) and Kayanuma (1988) the high-energy shift predicted by the variationally calculated values is much stronger than that revealed in the experiments. The main reason for the

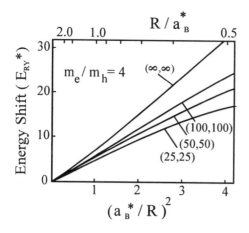

Fig. 3.4. Calculated high-energy shifts of the electron–hole-pair ground state with decreasing radius. In the brackets the height of the potential (V_e, V_h) is indicated in units of the Rydberg energy E_{Ry}. From Kayanuma and Momiji (1990)

difference is the use of the infinitely high barrier potential with vanishing wave functions at the boundary.

A further improvement of theory is consequently the consideration of finite barriers in the confining potentials $V_e(r_e)$ and $V_h(r_h)$. This approach has been used in a following work by Kayanuma and Momiji (1990). The overestimation of the confinement-induced blue shift could be decreased when using the finite barrier ansatz. The results of these calculations are shown in Fig. 3.4 for different values of the finite potential well V_e and V_h. The other parameters are adopted for CdS quantum dots with $m_h/m_e = 4$ for the mass ratio between hole and electron and $E_{Ry}^* = 30$ meV for the exciton binding energy. The energy gaps used for the calculations are $E_g^{glass} = 7$ eV, $E_g^{CdS} = 2.58$ eV and the band offsets are 50 %/50 %. The radius is given in units of the excitonic Bohr radius a_B.

The differences in the electron and hole effective masses inside and outside the dot has been additionally taken into account by Tran Thoai et al. (1990). Assuming $V_h = \infty$, but a finite barrier for the electrons, and an electron mass outside the dot between $m_e = m_0$ and $m_e = \infty$, depending on the mobility in the matrix, the quantum confined states have been calculated as functions of the dot radius. The theoretical result together with experimental data for small CdS quantum dots in colloidal solutions, taken from Weller et al. (1986), are shown in Fig. 3.5. For the liquid as the surrounding medium the electron is supposed to be mobile ($m_e = m_0$). A good agreement can be seen for $V_e = 40\,E_{Ry}$, corresponding to a potential height of about 1 eV for the liquid solution as the surrounding matrix material.

Exact wave functions can be obtained by extending the numerical methods towards matrix diagonalization techniques as done by Hu et al. (1990b) and Park et al. (1990). Compared with the results of variational methods, distinct differences in the one-pair energies were found, in particular in the range $R/a_B < 1$. Moreover, a change in the radial distribution has been

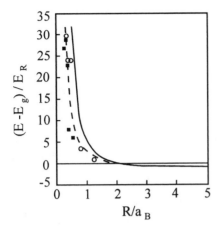

Fig. 3.5. Ground state energy for infinite well depth (solid line) and a depth of $V_e = 40\,E_{Ry}$ (dashed line) in comparison with experimental values [Weller et al. (1986)] of CdS quantum dots in colloidal solution. From Tran Thoai et al. (1990)

obtained for the calculated wave functions of the electron and hole as a consequence of the inclusion of the Coulomb interaction and the consideration of finite barriers. The Coulomb interaction between the electron and the hole perturbs the electron and hole wave functions, and the heavier particle is pushed towards the center of the dot. Since electron and hole can also penetrate into the barriers, the corresponding wave functions are no longer identical and finite transition probabilities appear for the dipole forbidden transitions. The inclusion of the Coulomb potential results in the correction of the confinement-induced energy shift to smaller values and to somewhat modified selection rules. As we will see later in Sect. 3.2, the efficiency of the matrix diagonalization technique becomes evident upon investigating the Coulomb interaction within a four-particle system.

Optical Transitions. Within the electron–hole-pair picture the absorption coefficient reads as

$$\alpha(\omega) \sim |\langle \Psi_{\text{pair}} | \mathbf{e} \cdot \hat{\mathbf{p}} | 0 \rangle|^2 \tag{3.22}$$

with $|0\rangle$ being the crystal ground state and $|\Psi_{\text{pair}}\rangle$ the excited pair state. There is an apparent analogy with the absorption process of excitons in bulk material. The transition probability for the exciton in the bulk is proportional to the probability of finding the electron and hole in the same unit cell of the crystal. The oscillator strength f_{Ex} is proportional to the volume in k-space necessary to form the exciton, i.e. the excitonic Bohr radius a_B,

$$f_{\text{Ex}} \sim \frac{|p_{\text{cv}}|^2}{\pi a_B^3}. \tag{3.23}$$

In quantum dots the transition probability is proportional to the spatial restriction of carrier motion in the quantum-dot volume due to the externally imposed quantum confinement. This has already been shown (3.16), describing the case of identical electron and hole wave functions without Coulomb interaction. Comparing the bulk exciton oscillator strength with that of electron–hole pairs in quantum dots, (3.18) and (3.23), we obtain

$$\frac{f_{\mathrm{QD}}}{f_{\mathrm{Ex}}} \sim \frac{a_{\mathrm{B}}^3}{R^3} , \qquad (3.24)$$

and an enhancement of the oscillator strength becomes evident when the size of the quantum dots decreases below the Bohr radius. Notice that this idealized consideration is valid only for the lowest $1s$-type transitions and not for the whole ensemble of optical transitions, nor in the case of different electron and hole wave functions, nor for wave functions extending into the barriers.

In this section the polarization energy due to the differences in the dielectric constants between the semiconductor quantum dot and the surrounding matrix were not considered and image force effects were neglected. Obviously, a more precise calculation can be made by including the image charges and induced polarizations $\delta V(r_{\mathrm{e}}, r_{\mathrm{h}}, \epsilon_1, \epsilon_2)$ in the Hamiltonian (3.19). It has been shown by Takagahara (1993a) that the surrounding medium with lower dielectric constant reduces the screening effect and thus enhances the stability of the pair state inside the dot. The result is an amplification of the exciton binding energy in quantum dots due to the dielectric confinement. For this topic we refer to Chap. 4.

3.1.3 Mixing of Hole States

To improve the models further, in the next step the description of the semiconductor band structure by simple parabolic bands has been dropped and more realistic bandstructures have been considered. For semiconductors like CdS, CdSe, ZnSe, etc., as well as most of the III–V compounds, the conduction band is formed from s-orbitals of the metal ions, whereas the valence band develops from p-orbitals of the S, Se or other group-V or group-VI elements.

Since the conduction band has its roots in s-type atomic orbitals, it causes only minor problems. For very small quantum dots, the nonparabolicity of the conduction band can become essential, since even the lowest states involve large k-wavenumbers [see, for example, the calculations made by Nomura and Kobayashi (1991)]. Most of the present theories approximate the conduction bands by simple parabolic bands. Treatments that go beyond these models often use the Kane-model (Kane 1957), suited both for conduction and valence band description.[1]

While the conduction band in most of the cases is well approximated by parabolic bands with only 2-fold spin degeneracy at $k = 0$, the valence band is not. In Fig. 3.6 the bulk bands are shown for the zinc-blende and wurtzite

---

[1] Kane (1957) developed a diagonalization technique which allows modeling of nonparabolicities of the bulk bands.

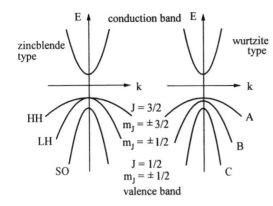

Fig. 3.6. Bandstructure of zinc-blende-type and wurtzite-type semiconductors

crystal structure. In zinc-blende-, i.e. T_d-type crystals, the top of the valence band at $k = 0$ is sixfold degenerate due to the p-type character of the atomic orbitals (omitting spin–orbit coupling). When we include the spin–orbit interaction, this leads to a reduction of the valence band degeneracy. The valence bands are then classified with respect to the total angular momentum J, representing the sum of the orbital angular momentum and the spin angular momentum.

Combining the orbital momentum 1 and the angular momentum 1/2 of the spin, one may construct a fourfold degenerate valence band with the total angular momentum $J = 3/2$ ($m_J = \pm 3/2; \pm 1/2$) and a twofold degenerate valence band with $J = 1/2$ ($m_J = \pm 1/2$). At $k = 0$ (the Γ-point of the Brillouin zone) the two bands $J = 3/2$ and $J = 1/2$ are split in energy with the energy separation given by the spin–orbit coupling constant Δ_{SO}. Depending on the semiconductor material, the top of the valence band is formed by the $J = 3/2$-states (as shown in Fig. 3.6 and typical for most of II–VI and III–V semiconductors) or by the $J = 1/2$-states (e.g. CuCl).

In bulk materials and quantum well structures the terms 'heavy-hole' (HH), and 'light-hole' (LH) subbands are adopted for the two uppermost valence bands and the term 'spin–orbit split-off band' (SO) for the lowest valence band.

Wurtzite-type crystals can be considered as a small perturbation of the T_d symmetry. In these crystals, also at $k = 0$, the degeneracy of the two uppermost valence bands is removed due to the crystal-field splitting. In bulk semiconductors of wurtzite-type, the three valence bands are denoted as A-, B- and C-band.

In the following, we will consider the more general case of zinc-blende-type semiconductors with the valence band top arising from the $J = 3/2$-states. The degeneracy of the m_J subbands is lifted for $k > 0$, i.e. away from the zone center. The effective masses of the $m_J = 3/2$ and the $m_J = 1/2$ bands (and by this the curvatures of these bands) are distinctly different. The dispersion of the hole energies is not parabolic nor isotropic, and the deviations are accounted for by the introduction of the Luttinger parameters $\gamma_1, \gamma_2, \gamma_3$ (Luttinger 1956) in the Hamiltonian for description of the hole masses.

In applying this approach to quantum dots, more complicate models for the kinetic energy term of the hole $\hat{H}_h$ have been introduced in the Hamilton operator $\hat{H}$ by, for example, Xia (1989), Grigoryan et al. (1990), Sercel and Vahala (1990), Koch et al. (1992), Nomura and Kobayashi (1992a), Efros (1992a), and Ekimov et al. (1993), treating the Hamiltonian

$$\hat{H} = -\frac{\hbar^2 \nabla_e^2}{2m_e} + \hat{H}_h - \frac{e^2}{\epsilon_2 |r_e - r_h|} + V_e(r_e) + V_h(r_h) . \tag{3.25}$$

As proposed by Luttinger (1956) the hole Hamiltonian $\hat{H}_h$ can be expressed for cubic materials with strong spin–orbit coupling by

$$\hat{H}_h = \hat{H}_{LU} = (\gamma_1 + \frac{5}{2}\gamma_2) \frac{\hat{p}^2}{2m_0} - \frac{\gamma_2}{m_0} (p_x^2 J_x^2 + p_y^2 J_y^2 + p_z^2 J_z^2)$$
$$- \frac{2\gamma_3}{m_0} [\{p_x p_y\}\{J_x J_y\} + \{p_y p_z\}\{J_y J_z\} + \{p_z p_x\}\{J_x J_y\}] \tag{3.26}$$

where m_0 is the free electron mass, γ_1, γ_2, and γ_3 are the parameters introduced to describe the valence-band dispersion, $\hat{p}$ is the hole momentum operator, and J is the 3/2-angular momentum operator.[2] Equation (3.26) was simplified by Baldereschi and Lipari (1971, 1973) by reducing the expression to spherical symmetry. Small contributions of terms of cubic or hexagonal symmetry were neglected and the parameter μ was introduced to give the strength of the spherical spin–orbit interaction. The Baldereschi–Lipari Hamiltonian $\hat{H}_{BL}$ is given by

$$\hat{H}_h = \hat{H}_{BL} = \frac{\gamma_1}{2m_0} \left[\hat{p}^2 - \frac{\mu}{9} (\mathbf{p}^{(2)} \mathbf{J}^{(2)}) \right] \tag{3.27}$$

with the coupling parameter μ defined by

$$\mu = \frac{6\gamma_3 + 4\gamma_2}{5\gamma_1} . \tag{3.28}$$

Here $\mathbf{p}^{(2)}$ and $\mathbf{J}^{(2)}$ are second-rank tensor operators, explained in detail in Baldereschi and Lipari (1971, 1973). The involvement of the full valence-band structure results in a qualitative new description of the electron and hole energy states, already in systems with only two-dimensional (2D) confinement. Whereas in bulk zinc-blende semiconductors, the uppermost valence band gives rise to *one* series of exciton states, the 2D-confinement lifts the degeneracy of the upper valence-band multiplet because the slab-shape symmetry is lower than T_d (for more details see Bastard 1988). Two series of quantized excitons, labelled heavy-hole (HH; $m_J = 3/2$) and light-hole (LH; $m_J = 1/2$) excitons, have been observed in 2D quantum wells and superlattices. In the direction of growth of the quantum well (usually

---

[2] $\{ab\} = (ab + ba)/2$

$z\|[001]$) one obtains for the effective masses of the two separate subbands $m_{z,\text{HH}} = m_0/(\gamma_1 - 2\gamma_2)$ and $m_{z,\text{LH}} = m_0/(\gamma_1 + 2\gamma_2)$, where the γ_i are the corresponding Luttinger parameters (Luttinger 1956).

As long as the spherical symmetry is retained in quantum dots, it is very unlikely that the m_J-degeneracy of the uppermost $J = 3/2$-valence band can be lifted (examples for a further splitting of states will be discussed in Sect. 3.1.4). However, a strong, confinement-induced valence-band mixing drastically changes the energies and wave functions, as we will see in the following.

For a spherical quantum well with an infinite potential barrier, Xia (1989) has calculated the Hamiltonian (3.27) with $\mu = 2\gamma_2/\gamma_1$, assuming only a small band warping ($\gamma_2 = \gamma_3$). He obtains the important result that the variation of the energy of the hole states is not monotonic with μ (ranging from 0 to 1) due to the mixing of heavy- and light-hole states by the confinement. The hole wave functions are now linear combinations of the different valence-band states and have mixed s–d-type symmetry. Similar to the in-plane motion in two-dimensional systems where the notations HH and LH are no longer striking, heavy-hole and light-hole states are no longer single eigenstates. The hole levels are now characterized by the quantum number of the total angular momentum $F = L + J$, which has to be conserved in optical transitions. Here L is the orbital angular momentum of the envelope function obtained from the confinement problem and J is the angular momentum from the Bloch-part of the wave function. The total quantum number F and the parity are preservation values. In addition to the minimum quantum number L, the orbital angular momentum contains the value $L + 2$ because (3.27) couples states with the same F and equal parity.

For the classification of the energy levels and optical transitions, the single particle quantum numbers for electron and holes from the particle-in-the-box problem [the former labelling according to nl, (3.8) and (3.9)], have to be replaced by suitable new notations. Usually the electrons are treated retaining the labels $1s$, $1p$, $1d$... for the $n = 0$ states and $1s$, $2s$, $3s$... for the $n \neq 0$ states, but supplied with the index 'e' for 'electron'. (Small letters are used to denote the true quantum numbers of single-particle wave functions and capital letters if relating to quantum numbers which are built up from combinations of different types of wave functions).

The notation of the hole levels has to provide the information about F and L and possibles terms for doing so could be

$$n^*(L, L+2)_F \tag{3.29}$$

with L and $L + 2$ the angular momentum arising from the involved χ_{nl} of the envelope function, $F = L + J$ the total angular momentum (containing the projection m_F from $-F$ to F), and n^* the counter for ground state and first, second, third etc. ... excited state. Then, for example, the letter S for the angular momentum involves $L = 0, 2$ and correspondingly P the values $L = 1, 3$ etc. .

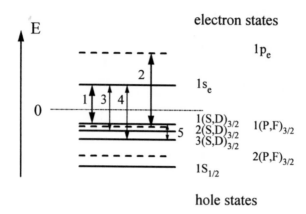

Fig. 3.7. Optical transitions of a quantum dot taking into account the mixing of the valence-band states.

In Fig. 3.7 we illustrate the energy-level scheme of the energetically lowest electron and hole states and the allowed optical transitions. States with an even quantum number L of the orbital angular momentum are plotted by solid lines and states with an odd L by dashed lines. The energetically lowest hole states arise from the $J = 3/2$ subband. The optical transitions are determined here from both the properties of the Bloch part and the envelope part of the wave functions. Supposing the survival of the basic properties obtained from the simplified particle-in-the-box-model (3.8, 3.9) one would expect that the first hole level with combinations of s- and d-type wave functions probably exhibits the $1s$-symmetry as the dominant part. The second and third hole levels contain correspondingly the $1p$ and $1d$ part etc. Therefore, the transitions 1 and 2 essentially correspond to the $1s_h \longrightarrow 1s_e$ and to the $1p_h \longrightarrow 1p_e$ transitions known from the simplified model. However, major changes should appear in the energy separation between the hole levels in comparison to the solution of (3.8) and (3.9). As we will see below, for special semiconductor materials the valence-band coupling results in a decrease in the level spacing of the former $1s$- and $1p$-hole levels.

In fact, the notation in Fig. 3.7 contains all wave functions involved, but is not very convenient for daily use. In the following we will use the notation proposed by Xia (1989) and by Baldereschi and Lipari (1973) with the labelling according to the *minimum* quantum number L involved in the wave function. (Another possibility often found in the literature is the use of the *dominant* contribution in L for abbreviation.)

The ground state of the hole with $n = 1$, $F = 3/2$ and $L = 0, 2$ is then given by $1S_{3/2}$, and the first electron–hole transition is $1S_{3/2} \longrightarrow 1s_e$ (transition 1 in Fig. 3.7). Furthermore, the transition $1P_{3/2} \longrightarrow 1p_e$ (transition 2 in Fig. 3.7) is dipole-allowed. Due to valence-band mixing of wave functions with s- and d-type symmetry and due to the Coulomb interaction, former for-

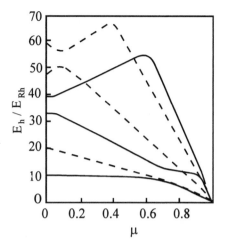

Fig. 3.8. Calculated energies of the hole states $S_{3/2}$ (solid lines) and $P_{3/2}$ as function of the coupling constant μ, from Koch et al. (1992). For other parameters see text

bidden optical transitions with $\Delta n \neq 0$ such as the transitions $2S_{3/2} \longrightarrow 1s_e$ and $3S_{3/2} \longrightarrow 1s_e$ (3 and 4 in Fig. 3.7) acquire significant oscillator strength and appear in the linear and nonlinear-optical spectra. The difference in the overlap integrals of the wave functions is only a factor of two for the transition from the $1S_{3/2}$- and the $2S_{3/2}$-hole state to the $1s_e$-electron state according to the calculation made by Xia (1989) using $\mu \sim 0.7$ for the spin–orbit coupling parameter. Transition 5 in Fig. 3.7 reminds us that 'intraband' transitions also take place within the ladder of either electron or hole states.

The state $1S_{1/2}(C)$ has been added to the scheme to recall that an additional transition from the split-off valence-band states (corresponding to the C-band in bulk wurtzite II–VI materials) could be expected, in particular in semiconductors with small energy splitting Δ_{SO}. Therefore, in CdS and CdS_xSe_{1-x} mixed crystals, the situation can become even more complicated due to the similar order of magnitude of the confinement-induced energy shifts and the value of Δ_{SO}.

The present theoretical work is aimed at calculating the exact eigenvalues and their energy separations as a function of dot sizes, as well as revealing the dominant symmetry of the wave functions that determine the selection rules for optical transitions. Besides the approaches based on the effective mass approximation and outlined as follows, the detailed valence-band structure has likewise been considered in tight-binding calculations published by Einevoll (1992) and Ramaniah (1993).

The confinement-induced valence-band mixing has been investigated by Koch et al. (1992) by combining the Hamiltonian $\hat{H}_{BL}$ (3.27) with the numerical matrix diagonalization method by additionally including the Coulomb interaction and by using the parameter μ according to (3.28). Most of the II–VI semiconductors considered here possess values of the coupling parameter μ around 0.7 [CdS 0.75; ZnS 0.751; ZnSe 0.795; ZnTe 0.755; CdTe

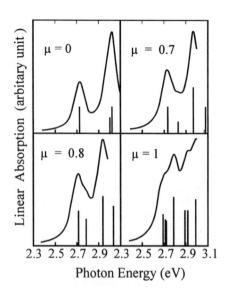

Fig. 3.9. Calculated linear absorption spectra of CdS quantum dots ($R = a_B$) with the valence-band coupling constant μ as parameter. From Koch et al. (1992)

0.855 (Baldereschi and Lipari 1973)]. Therefore, the valence-band mixing is not negligible and leads to modifications of the hole energies, the electron–hole-pair wavefunction and the transition probabilities. In Fig. 3.8 the lowest quantum-confined hole levels $S_{3/2}$ and $P_{3/2}$ have been plotted as a function of the band coupling parameter μ. In the case $\mu = 0$ there is no coupling, and solutions of the parabolic band approximation were obtained, which remain a good description up to $\mu < 0.3$. Near degeneracy of the two lowest hole states has been obtained for $\mu \sim 0.7$ as a peculiarity of the confinement-induced valence-band mixing. For that reason in Fig. 3.7 the $1P_{3/2}$-level is plotted close to the $1S_{3/2}$-level. The degeneracy of the two lowest hole states, $1P_{3/2}$ and $1S_{3/2}$, has been proved in two-photon absorption experiments (TPA) by Kang et al. (1992) for small CdS quantum dots. The one- and two-photon absorption peaks have been found to nearly coincide in energy according to the degeneracy of the lowest $S_{3/2}$- and $P_{3/2}$-states in semiconductor dots with $\mu \sim 0.7$.

Figure 3.9 from Koch et al. (1992), shows examples of calculated linear quantum-dot absorption spectra with the valence-coupling strength μ as the parameter. The other parameters have been chosen from CdS ($E_g = 2.58$ eV, $E_{Ry}^{ex} = 27$ meV, $R = a_B$, $m_e \gamma_1/m_0 = 0.5$, with a homogeneous broadening of $2E_{Ry}$). Due to the opening of new dipole-allowed transitions, a shoulder appears in the linear absorption spectra on the high-energy side of the lowest state. The inserted lines indicate the oscillator strengths, which are very sensitive with respect to the hole-coupling constant μ. As a further result of the analysis of Koch et al. (1992) it has been found that the coupling between the different valence bands modifies not only the energy states. Moreover the radial distribution of the electron and hole wave functions drastically

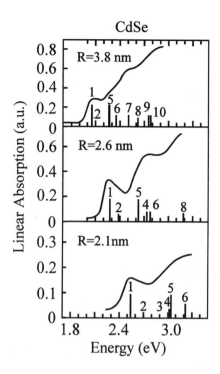

Fig. 3.10. Linear absorption spectra and calculated positions of the electron–hole pair transitions for different sizes of CdSe quantum dots. Data taken from Ekimov et al. (1993)

changes compared to the calculation without valence-band mixing effects. The maximum of the hole radial probability density differs considerably from that of the electron. The band coupling effect causes a polarization inside the quantum dot, which might be one more reason for modifications of the electron–phonon coupling strength, as will be discussed in Chap. 5.

The widely investigated system of CdSe quantum dots has been treated in detail by Ekimov et al. (1993) and Efros (1992a). The Coulomb interaction between electrons and holes, the valence-band degeneracy and the nonparabolicity of the conduction band have been considered. The absorption spectra were calculated for CdSe quantum dots and compared with the linear absorption measured at high-quality samples of CdSe-doped glasses. The results are shown in Fig. 3.10. The structures found in the absorption spectra were assigned to the calculated energies and oscillator strengths of the lowest optical transitions allowed (see also Table 3.1). It was confirmed that the hole-energy levels are apparently different from those obtained by simple parabolic-band models. The energetic difference of the lowest states increases, but by no means can it be explained by a $1/R^2$-dependence with decreasing size. The first two absorption peaks were assigned for CdSe quantum dots to the $(1S_{3/2}, 1s_e)$ and the $(2S_{3/2}, 1s_e)$ 1EHP states in the range of sizes between $R = 2$ to $4\,\mathrm{nm}$. The calculated order of the lowest optical transitions can be taken from Table 3.1.

Table 3.1. Order of the first electron–hole-pair transitions calculated for CdSe quantum dots with $R < a_B$, after Ekimov et al. (1993). $nP^l_{1/2}$ is the level of the light hole, $nP^{so}_{1/2}$ the level of the spin–orbit split-off band

No.	Transition	No.	Transition
1	$1\,S_{3/2} - 1\,s_e$	6	$1\,P^l_{1/2} - 1\,p_e$
2	$2\,S_{3/2} - 1\,s_e$	7	$3\,S_{1/2} - 1\,s_e$
3	$1\,S_{1/2} - 1\,s_e$	8	$2\,S_{3/2} - 2\,s_e$
4	$2\,S_{1/2} - 1\,s_e$	9	$1\,P^{SO}_{1/2} - 1\,p_e$
5	$1\,P_{3/2} - 1\,p_e$	10	$4\,S_{3/2} - 2\,s_e$

The published results show that the hole-energy states are very sensitive to the choice of the parameter μ for the description of the valence band coupling ($\mu = 2\gamma_2/\gamma_1$, $\gamma_2 = \gamma_3 = \gamma$, in spherical approximation). For example, to explain the linear spectra, the values $\gamma_1 = 2.1$ and $\gamma = 0.55$ have been used by Ekimov et al. (1993) yielding $\mu = 0.52$. To explain magneto-optical properties, $\gamma_1 = 1.27$ and $\gamma = 0.34$ with $\mu = 0.538$ has been applied by Nomura et al. (1994). To calculate biexciton binding energies, the value $\mu \approx 0.7$ has been introduced in the numerical computations by Hu et al. (1990a).

To test the sensitivity of the energies of the lowest 1EHP transitions to changes in the valence band coupling strength, the transition energies have been computed following the model published by Xia (1989) and Grigoryan et al. (1990) and plotted in Fig. 3.11. For the calculations, a few basic assumptions have been made: The quantum dots are considered to be ideal spheres with zinc-blende crystal structure. The Coulomb interaction between electron and hole is treated in the frame of perturbation theory. Effects of dielectric confinement are neglected. For the electron states, basically parabolic bands are assumed. They include, however, higher conduction bands. For the hole states the Luttinger Hamiltonian in spherical approximation is used (Luttinger 1956, Baldereschi and Lipari 1971, 1973). The potential barriers for the electrons are considered to be finite, but very high, the barriers for the holes are considered to be infinite. These assumptions are the usual approximations in the present theoretical work (see for example Xia 1989, Ekimov et al. 1993).

As seen in Table 3.1, the lowest four optical transitions contain the same $1s_e$-electron level. To find a representation that fits the theoretical curves to the experimental data, the energy separation ΔE of the higher one-pair states with respect to the lowest one-pair state ($1S_{3/2}, 1s_e$) is plotted, and not the absolute energy of the corresponding transition. Such a plot was introduced

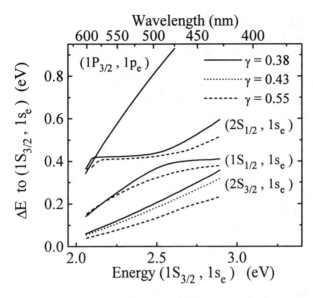

Fig. 3.11. Energy difference ΔE between the lowest optical transitions calculated for various values of $\gamma = 0.55, 0.43, 0.38$; for other parameters see text. From Woggon et al. (1996b)

by Norris et al. (1994) to normalize the electron contribution to the optical transition energies. Furthermore, for the x-axis the energy of the lowest one-pair transition instead of the dot radius has been used to further minimize uncertainties caused by the experimental determination of sizes. That energy value is easily accessible by the experimental conditions, such as, for example, the pump energy in nonlinear absorption experiments or the detection energy in photoluminescence excitation spectroscopy.

For the results presented in Fig. 3.11, the parameter γ has been varied as $\gamma = 0.55, 0.43$, and 0.38 corresponding to $\mu = 0.52, 0.41$ and 0.36, respectively. The other parameters are $\gamma_1 = 2.1$, $m_e = 0.11\,m_0$ for the electron mass, $E_G = 1.84\,\text{eV}$ for the energy gap, and $\Delta_{SO} = 0.42\,\text{eV}$ for the spin–orbit splitting energy. From the curves in Fig. 3.11 we can derive the following: (i) The radius dependence of the energy states is strongly influenced by γ. (ii) For the whole size range the two lowest transitions are those to the $(1S_{3/2}, 1s_e)$ and $(2S_{3/2}, 1s_e)$ states, i.e. the next excited transition is the one involving the $2S_{3/2}$ hole and the $1s_e$ electron state. (iii) The transition to the $(1P_{3/2}, 1p_e)$ state always appears above the $(1S_{1/2}, 1s_e)$ state in the calculations. However, the treatment of the p-electrons is very crude in this approximation and the energy of that transition has the largest error. As we will see below, the very large energy separation of the $(1P_{3/2}, 1p_e)$-state to the ground state is not very likely for smaller dot sizes. (iv) Level crossing is avoided between the $(1S_{1/2}, 1s_e)$ and the $(2S_{1/2}, 1s_e)$ states. The avoided level crossing is an important result and connected with the mixing of L and $L+2$ states in the hole wave functions. For the $S_{1/2}$ hole states the quantum number $F = 1/2$ can be composed both from d- and s-like states with $J = 3/2$, $L = 2$, and $J = 1/2$, $L = 0$. In this case, the s- and d-contributions can arise from different valence bands. When valence-band mixing effects are absent, the transitions from the $J = 3/2$ hole state and from the $J = 1/2$ hole state of the split-off band have no common set of basis wave functions and the two curves may intersect in their size dependence. However, including the valence-band mixing effects the two states have contributions both from s- and d-type wave functions and differ only in the relative weights of these wave functions. Therefore, intersections are forbidden and level crossing is avoided in the spectra. The oscillator strength of transitions starting from the $S_{1/2}$ hole states can strongly vary when the size of the dots changes. In any case, it will be weaker than that of the neighboring strongly allowed transition to the $(1P_{3/2}, 1p_e)$ state.

The exactness of the model is softened for sizes where the effective mass approximation is no longer applicable and expected here for radii below ≈ 1.2 nm, i.e. above energies of ≈ 2.9 eV. Hence, only dot radii that are far from cluster physics are relevant for the comparison between theory and experiment. For an estimation of the dot radii the following values can be used: an energy of 2 eV corresponds to sizes of $R \approx 3.5$ nm, 2.4 eV to $R \approx 1.8$ nm, and 2.8 eV to $R \approx 1.3$ nm.

All the considerations above are limited to crystals exhibiting the zinc-blende-type lattice structure. Extending the calculation to wurtzite-type symmetry (Efros 1992a,b), further modification could be found compared with the results from cubic symmetry. That topic will be the subject of the next section.

3.1.4 Splitting of States

In quantum dots of zinc-blende type semiconductors, as considered above, the energetically lowest state is a $(1S_{3/2}, 1s_e)$-pair state with the hole arising from the uppermost $J = 3/2$ valence band and the electron from the lowest conduction band with the total angular momentum only determined by the spin quantum number $\pm 1/2$. The pair-binding energy is primarily determined by the size of the dot, the height of the potential barrier and the strength of the Coulomb interaction. This pair state has an analogy in the exciton state of the bulk material. The bulk exciton is likewise composed of an s-type conduction band electron and of a $J = 3/2$ hole. The exciton state in the bulk is 8-fold degenerate at $\boldsymbol{k} = 0$. This degeneracy can be lifted by the crystal field, for example in wurtzite-type semiconductors, or by exchange interaction. In bulk material, the exchange interaction results in an internal structure of the excitonic states that contains both optically allowed and forbidden transitions. The exchange interaction comprises a long- and a short-range contribution (for details see, for example, Onodera and Toyozawa 1967; Cho 1976; Rössler and Trebbin 1981; Bassani et al. 1993; Andreani and

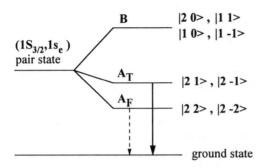

Fig. 3.12. A simplified scheme illustrating the lifting of degeneracy of the $(1S_{3/2}, 1s_e)$ pair state due to nonspherical shape, crystal-field splitting and exchange interaction. The notation of the states is $|N, N_m\rangle$ (see text)

Bassani 1990; Jorda et al. 1992). The strength of both the short-range and the long-range part is proportional to the spatial overlap of the electron and hole wave functions. Whereas only the excitonic wave functions of essentially the same unit cell contribute to the short-range part, the long-range part contains the dipolar interactions between several lattice cells within a distance comparable to the Bohr radius. In bulk semiconductors the long-range exchange interaction causes the longitudinal–transversal splitting of the exciton energy. This splitting energy $\Delta\omega_{\text{LT}}$ is proportional to the coupling strength of the exciton to the photon.

The origin of the short-range part has a close similarity to the singlet–triplet splitting in two-electron systems. Fermions with equal spin quantum number have to be different in their spatial coordinates to fulfill the Pauli-principle. When the two particles are forced to be at the same place, the system reacts with an energy splitting. If the energetically lowest state is the triplet state, optical transitions are prohibited and the luminescence is characterized by a long lifetime. Since both the long-range and short-range part of the exchange interaction are proportional to the spatial overlap of the electron and hole wave functions, a strong enhancement is expected in systems of lower dimensionality. For quantum dots, the problem of internal structure of the lowest pair states has been theoretically treated by Efros (1992a), Efros and Rodina (1993), Takagahara (1993a), Nomura et al. (1994), Romestain and Fishman (1994), Nash et al. (1995), and Nirmal et al. (1995). The lifting of degeneracy of the lowest pair-states is very likely in quantum dots since, for example, deviations from spherical shape, presence of wurtzite-type lattices etc. have been observed experimentally. In small quantum dots with radii below the bulk excitonic Bohr radius an enhancement of the exchange interaction proportional to Ω_0/R^3 has been predicted, with Ω_0 the volume of the unit cell and R the radius of the dot. For sizes with $R/a_{\text{B}} < 0.5$ the splitting energy can be enhanced very much over the bulk value [up to two orders of magnitude (Takagahara 1993)]. Quantum dots of semiconductors with large Bohr radius, like e.g. GaAs, should exhibit a more pronounced fine structure of the pair states than those from materials where the Bohr radius nearly equals the lattice constant, such as, for example, CuCl.

As in bulk material, the theoretical description starts from an 8-fold degenerate pair state (Fig. 3.12). The electron is only characterized by its spin quantum number $s = 1/2$ with the projections $m_s = \pm 1/2$. The hole states are described by the quantum number $F = L + J$ due to valence-band mixing. For the $1S_{3/2}$-hole state, $F = 3/2$ holds with the projections $m_F = \pm 3/2, \pm 1/2$. To characterize the pair state $(1S_{3/2}, 1s_e)$, the total pair angular momentum $N = F + s$ with its projection N_m has been introduced. The eight single states differ in their quantum numbers according to all possible combinations $|N, N_m\rangle = |F, s, m_F, m_s\rangle$. The presently used notations for the 8 states are 1^U for the states $|1, 1\rangle$ and $|1, -1\rangle$, 0^U for the state $|1, 0\rangle$, 1^L for the states $|2, 1\rangle$ and $|2, -1\rangle$, 0^L for the state $|2, 0\rangle$, and 2 for the states $|2, 2\rangle$ and $|2, -2\rangle$ (Nirmal et al. 1995).

The crystal field and nonspherical shape both lift the degeneracy and split the state into two groups labelled A and B in Fig. 3.12. Calculations made by Nirmal et al. (1995) have shown that the group B energy levels undergo further significant energy splitting in the limit of strong nonsphericity and for dot radii below 1.5 nm. In the case of near-spherical nanocrystals however, with dot radii between 1.5 and 3 nm, the group A and B levels are well-separated in energy by a few tens of meV. In the following we will concentrate on the group A levels. Due to exchange interaction, the A-states split further into two pair states with the projections $N_m = \pm 1$ and ± 2 of the pair total angular momentum $N = 2$. The lowest state is labelled A_F and is dipole forbidden because light with angular momentum 1 couples to a state with $N_m = \pm 2$ only in higher orders of perturbation theory. Thus the energetically lowest state in Fig. 3.12 absorbs only weakly but can emit after relaxation from higher excited states. The pair state with $N_m = \pm 1$, here labelled A_T, strongly couples to the radiation field.

In most of the calculations the long-range part, the dipole–dipole contribution to the exchange splitting has been neglected and it has been argued that (i) the quantum dots are too small to couple efficiently to the radiation field[3], and (ii) the lowest state has spherical symmetry and therefore its contribution vanishes after integration over the angular coordinates. However, the neglect of the long-range part in the exchange interaction seems to be a very crude approximation. First proposals for the contribution of the long-range exchange interaction to the splitting energy in quantum dots have been made by Takagahara (1993a). A non-vanishing long-range exchange energy was expected for the case of nonspherical shape of the quantum dots and if the envelope function had components with higher l quantum numbers. For crystallites with large bulk values of the splitting energy $\Delta \hbar \omega_{LT}$, the value

---

[3] In quantum dots the translational symmetry is limited and therefore the concept of center-of-mass motion with a K-vector coupling to the radiation field is rarely discussed.

of the long-range part could be of similar quantity to the short-range part of the exchange interaction and, consequently, might also influence the fine structure. The pair states have finite oscillator strength and even small quantum dots contain more than 1000 elementary cells. Additionally, the samples usually have thicknesses of about 100 μm and volume fractions of semiconductor material of ~ 1 %, resulting in a huge number of dipoles. Hence, the coupling to the radiation field should have a nonzero, finite strength also in quantum dots. Thus, the question of the long-range exchange interaction in zero-dimensional quantum structures remains a subject for further investigations.

Experimental evidence for the splitting of the electron–hole-pair states in quantum dots was obtained by Chamarro et al. (1995, 1996) in strongly confined CdSe quantum dots. The authors observed a size-dependent energy gap between absorption and luminescence ranging from 1 to 10 meV. They explained the shift between the absorption and luminescence peak by exchange splitting of the $(1S_{3/2}, 1s_e)$ pair state with the lowest state being a forbidden state. As a further argument, they mentioned the time-dependent change of the polarization degree of the luminescence. Five exciton states produced by exchange interaction and nonspherical symmetry were introduced by Nirmal et al. (1995) for the splitted lowest 1EHP state in CdSe quantum dots. By means of magnetic field experiments, the transition probability could be tuned between the allowed transition with total angular momentum $N = 1$ and the energetically lower forbidden transition with total angular momentum $N = 2$. Obviously, this property is introduced by the spin-part of the wave functions, reflected by its sensitivity to the external magnetic field. Further experimental examples for the fine structure of the lowest electron–hole-pair state will be given in Sect. 3.1.6. Currently, for silicon-based nanostructures, the contribution of singlet–triplet splitting to the luminescence decay time is under very intensiv discussion (Calcott et al. 1993a,b; Romestain and Fishman 1994; Nash et al. 1995) and we will return to this topic in Chap. 9.

3.1.5 Indirect-type Quantum Dots

Indirect-gap properties of semiconductor quantum structures may arise both from the difference in k-space and in r-space between the lowest electron and hole-wave functions. The different types of indirect-gap structures are depicted in Fig. 3.13. The occurrence of quantum confinement in small artificial structures of the indirect-gap semiconductors silicium and germanium has been reported as well as the arrangement of different semiconductor materials in dot-like structures with a succession of the band offsets similar to type-II semiconductor quantum wells. The former will be treated in more detail in Chap. 9, the latter in Chap. 8.

Quantum structures manifest their indirect character mainly by slow recombination rates in the optical transitions. Hence, the design of indirect-gap

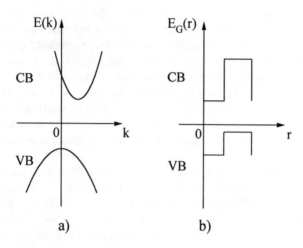

Fig. 3.13. Indirect-gap quantum dots with respect to (a) momentum space and (b) coordinate space

quantum structures aims to engineer the radiative transition probabilities and oscillator strengths. The optical transitions in indirect-gap bulk semiconductors require the participation of phonons to fulfill the k-selection rules. For quantum structures of semiconductors with indirect-gap band structure, the supposition is that the zero-phonon transition probability might be enhanced as a result of the mixing of momentum states from both the envelope and the Bloch part of the wave functions.

The question of whether the oscillator strength increases in indirect-gap semiconductors when the dimensionality decreases has been investigated by Tolbert et al. (1994a), and Tolbert and Alivisatos (1994b) on CdSe quantum dots. In contrast with the group IV elements Si and Ge, the material CdSe exhibits both direct and indirect band structure. A pressure-induced solid–solid phase transition between wurtzite structure (direct-gap) to rock salt structure (indirect-gap) occurs in the bulk semiconductor. Around a pressure of 3 GPa the CdSe bulk material can be reversibly converted from the wurtzite-lattice to the rock-salt lattice. The optical transition energy of the rock-salt modification is in the infrared spectral range at 0.67 eV. This peculiarity provides the possibility of direct examination of changes in the optical transition probability if going from direct to indirect band structure, however in nanocrystals of the *same* size. Tolbert and Alivisatos (1994b) showed the presence of the structural phase transition for nanocrystals of CdSe with radii between 1 nm and 1.7 nm using TEM, X-ray diffraction and SAXS measurements. Simultaneously, the oscillator strengths have been studied in optical experiments. Since in luminescence the intrinsic electronic transition probability is often influenced by trapping effects, the absorption tails have been investigated and compared for both modifications. In wurtzite-type quantum dots, the absorption features arising from the quantum confinement are well observable. However, this sign for quantum confinement disappears at the structural phase transition pressure (≈ 7 GPa) and the absorption spectrum becomes rather smooth with a gradual slope towards lower energies. Within the experimental error, no variation in the ratio of direct to indirect gap absorption intensity could be observed in the size range between $R = 0.96$ nm and $R = 1.73$ nm. Therefore the results can be summarized in the statement that the change in oscillator strength only weakens the indirect-gap character but cannot result in a dominance of the direct-gap properties.

A further kind of indirect-type quantum structures is based on different spatial distributions of the electron and hole wave functions, i.e. the electron and hole reside in different materials. In a so-called type-II quantum dot, one of the particles (i.e. the electron) is spatially confined in the dot material, whereas the other particle (hole) is situated in the barrier (matrix) material (Fig. 3.13). The Hamiltonian describing this problem is essentially the same as in (3.19) despite the explicit form of the confinement potentials V_e and V_h. For the type-II character it is necessary that V_e and V_h at the dot boundary show the same functional dependence on the spatial coordinate, i.e. either both increase or both decrease with r. The energy states as the solutions of the Hamiltonian (3.19) with the modified expressions for the potential barriers V_e and V_h reflect the interplay between the attracting Coulomb interaction and the separating potential barriers.

The binding energy of an electron–hole pair in type-II quantum dots has been calculated by Laheld et al. (1993) and Rorison (1993) for infinite potential barriers and later by Laheld et al. (1995) for the problem with finite barriers. Laheld et al. (1993) discuss both analytical approximations for the limiting cases of small and large dot radii and carry out variational calculations. In the case of very small dot radii, the problem resembles the textbook problem of finding the binding energy of a hole in the field of an acceptor represented by an electron in its discrete ground state, despite the boundary condition at $|r_h| = R$ for the wave functions. For dot sizes larger than the Bohr radius the result is strongly influenced by the ratio of electron and hole masses. A main result of the calculations is the tendency towards smaller electron–hole-pair binding energies in type-II structures compared with type-I structures.

The extension to finite barriers by Laheld et al. (1995) highlights the more interesting question of transition probabilites, since now the overlap of the electron and hole wave functions is sensitive to the band offset of the barrier material. The particle wave functions are able to penetrate the dot boundary. With decreasing size/potential height the electron wave function enters the barrier more and more and overlaps with the hole state, thus gradually increasing the oscillator strength. From the calculations by Laheld et al. (1995) a quadratic increase of the oscillator strength has been obtained when decreasing the radius R. For the binding energy, one obtains a similar result to that known for two-dimensional superlattices, namely a rise in the binding energy with decreasing dot radius to a maximum value followed by a sharp decrease below $R < a_B$. To explain this behavior, the idea of a smearing out of the charge distribution created by the confined particle at very small radii has been introduced. The maximum value of the binding energy then corresponds to the minimum extension of the confined particle wave function.

3.1.6 Experiments to Identify One-Pair States

Despite the theoretical progress, the present experimental data are strongly influenced by the real structure and quality of the samples investigated. Considering the whole ensemble of quantum dots, the theoretically predicted energy states are unlikely to resolve due to the inevitable distribution in dot sizes. Sometimes the linear absorption of, for example, II–VI nanocrystals in glass or polymer matrices shows no structures at all, even at very low temperatures. To reveal the individual energy states inherent to a single quantum dot, more sophisticated experimental techniques have to be applied. Unlike quantum wells or wires, quantum dots embedded in isolating matrices can hardly be studied by means of electric techniques. Contactless optical methods are used to probe the electronic properties of these structures. In high-quality samples of small dot sizes, substructures can be resolved by examination of the higher derivatives in the optical absorption spectra (Ekimov et al. 1993; Katsikas et al. 1990) or the application of modulation spectroscopy, for example, via an external electric field (Nomura and Kobayashi 1990; Sekikawa et al. 1992). However, these techniques are basically not size-sensitive and provide the data relevant to the mean size $\overline{R}$ of the ensemble only. To study size-dependent properties, a more effective approach is based on size-selective excitation using narrow-band tunable lasers. By this technique either photoluminescence or nonlinear absorption can be analyzed. Since most of the II-VI quantum dot structures, especially the ones in glass matrices, exhibit low quantum yield of the edge emission, nonlinear absorption sometimes can be considered as more suitable. Currently, single-dot spectroscopy by use of techniques providing high spatial resolution is going to become an efficient tool to monitor the individual dot properties, in particular to study epitaxially grown quantum structures (for this topic see Chap. 8).

Over long time periods, the published results from quantum dots were very much different for embedding either in glassy or organic matrices and the question arose as to whether quantum dots can have systematic properties independent of their preparation prehistory at all. In the following, the existence of a general size-dependence of the electronic properties will be tested for the case of CdSe quantum dots embedded in various matrix materials. We concentrate on the determination of the energy separation ΔE for the lowest few one-pair transitions. For that purpose we investigate a large variety of samples from different sources and manufacturing processes.

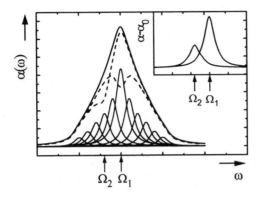

Fig. 3.14. Saturation of Lorentzian lines in the case of dominant inhomogeneous broadening within the two-level model for two different pump energies Ω_1 and Ω_2

The experimental data will be compared with present theoretical work about energy states (see Figs. 3.10 and 3.11), but for now effects of fine structure will be omitted.

To derive the information about the quantum dot energy states photoluminescence (PL), photoluminescence excitation spectroscopy (PLE), and differential absorption (PP) have been applied. All three experimental methods achieve size selectivity by using a spectrally narrow pump beam for the excitation. Furthermore, in PLE a narrow-band detection has been used in the low-energy tail of the absorption spectrum. Tuning the excitation through the whole absorption spectrum and assuming fast relaxation, the PLE probes the density of states. It is a probe for the discrete, absorbing electron–hole-pair states. The prerequisite for this experiment is an efficient 'band-edge'-related luminescence of the CdSe quantum dots as observed, for example, by Hoheisel et al. (1994), Norris et al. (1994), Rodrigues et al. (1995), and Wind et al. (1996).

In a pump-probe configuration (PP) the pump-beam excites resonantly a single quantum dot transition, whereas the weak, spectrally broad probe-beam simultaneously tests the whole absorption spectrum. By means of this so-called 'spectral hole burning', i.e. the measurement of the spectrally resolved change in the absorption coefficient $\Delta\alpha$ at selective laser excitation, the electronic states of a single quantum dot size can be revealed within a strong inhomogeneously broadened absorption band. In Fig. 3.14 an inhomogeneously broadened ensemble of two-level particles is sketched with resonant absorption at $\hbar\Omega_i$. Under a steady-state monochromatic excitation at $\hbar\omega_{\mathrm{pump}} = \hbar\Omega_1$, one part of the particles experiences a transition into the excited state. With increasing intensity I, the absorption coefficient $\alpha(I)$ decreases to the limit $\alpha(I) \to 0$ for $I \to \infty$. The product $\alpha(I)I$, i.e. the absorbed power, saturates. In the case of constant life time T_1 the absorption saturation can be described as

$$\alpha\left(I\right) = \frac{\alpha_0}{1 + I/I_{\mathrm{sat}}}, \tag{3.30}$$

where I_{sat} is determined by the probabilities of spontaneous and stimulated transitions (see Bloembergen 1977; Letokhov and Chebotaev 1977;
Demtröder 1991 for details). In a strong inhomogeneously broadened system,
present in quantum dots by a wide distribution of sizes, the maximum change
in the absorption coefficient α is correlated with the spectral position of the
pump laser. The single, homogeneously broadened Lorentzian line saturates
uniformly over all frequencies. The width of the 'burnt' hole is connected with
the homogeneous line-width Γ of the Lorentzian and can be exactly evaluated from the change $\Delta\alpha(\omega)$ if the saturation strictly follows the two-level model and, moreover, if the size distribution $P(R)$ and the energy dependence
$E_{QD}(R)$ are known (Bloembergen 1977; Shen 1984)[4]. In systems involving
more than two levels, the spectral width of the bleaching spectrum can be
used (at not too high excitation intensities) as a rough estimate for the homogeneous line-width Γ. If the system has excited states, and if transitions with
common initial or final states are allowed, a population or depopulation of
this state should result in a reduced absorption for all transitions coupled to
this state. Thus, according to the scheme presented in Fig. 3.7, a multi-band
bleaching under selective excitation is expected.

It should be mentioned that the spectra of differential absorption of small
quantum dots can become highly sensitive with respect to laser exposure.
Besides the transient spectral burning as a pure population effect, persistent
hole burning phenomena have also been reported (see Chap. 5).

In Fig. 3.15 the PL, PP and PLE data have been plotted for an ensemble
of CdSe quantum dots with an average size of $\bar{R} = 2.5$ nm and a deviation of
$\Delta R = \pm 10\%$. The detection energy in the PLE measurements and the pump
energy in the PL and PP experiments were chosen at 2.18 eV in the tail of the
absorption spectrum. At that energies, predominantly nanocrystals with sizes
larger than the average radius $\bar{R}$ are tested. From Fig. 3.15 it can be seen that
the photoluminescence (PL) itself is still characterized by a superposition of
states forming a broad band shifted to lower energies with respect to the
excitation. The degree of information is still low because neither the single
luminescent states nor possible phonon replica could be resolved.

The PP spectrum of Fig. 3.15 shows the absorption bleaching for dominant excitation resonant to the $(1S_{3/2}, 1s_e)$ 1EHP state. All s-like excited hole states contribute to the nonlinear absorption after the $1s_e$-electron
state is optically populated. The two structures in the bleaching spectrum at
2.19 eV and 2.28 eV are therefore attributed to the transitions to the lowest
$(1S_{3/2}, 1s_e)$ and $(2S_{3/2}, 1s_e)$ states. The $\Delta\alpha d$ signal around 2.61 eV has been
assigned to a transition with dominant $S_{1/2}$-symmetry from the spin-orbit
split-off valence band. The disadvantage of the PP measurement lies in the

---

[4] At very high excitations, hole burning within an inhomogeneously broadened
system is characterized by power-broadening effects in the hole widths according
to $\Gamma(I) = \Gamma_0 \left(1 + \sqrt{1 + I/I_{sat}}\right)$. In this case, the absorption at $\hbar\omega_{pump}$ is given
as $\alpha(I) = \alpha_0 (\omega_{pump}) / \sqrt{1 + I/I_{sat}}$ (Letokhov and Chebotaev 1977).

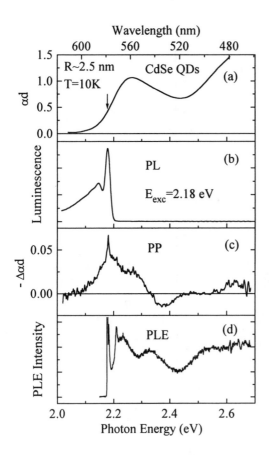

Fig. 3.15. (a) Linear absorption, (b) photoluminescence (PL) (the strongest peak indicates the excitation energy by the scattered pump light), (c) differential absorption (PP), and (d) photoluminescence excitation (PLE) spectra of CdSe quantum dots ($\bar{R} = 2.5\,\text{nm}$). The pump/detection energy is 2.18 eV. From Woggon et al. (1996b)

appearance of an induced absorption feature (for example around 2.37 eV) which is caused by the absorption of one pump- and one probe-photon and the formation of a two-pair state (see Sect. 3.2). This process has to be taken into account when fitting the real line positions. The narrow component around the pump is not an artefact, for example, from scattered pump light. It was first reported by Norris et al. (1995) for the differential absorption spectra of CdSe quantum dots. The narrow bleaching signal follows the pump detuning and rides in the $\Delta\alpha d$ spectra over a broad bleaching. When we deduce the single-dot absorption from the $\Delta\alpha d$ spectra, consistency with the experimentally measured line shape of the burnt hole could only be obtained by introducing a fine structure within the lowest ($1S_{3/2}, 1s_e$) pair state (Norris et al. 1995 and see below).

The PLE experiment shown in Fig. 3.15 exhibits the highest accuracy in resolving substructures within the lowest electron–hole-pair state, as can be seen in the narrow peak near the detection energy and the peak separated by the LO-phonon energy. The two main structures again were attributed to the transitions to the lowest ($1S_{3/2}, 1s_e$) and ($2S_{3/2}, 1s_e$) states. However,

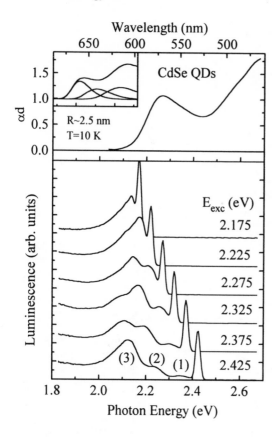

Fig. 3.16. Linear absorption and size-selective photoluminescence spectra of CdSe quantum dots ($\bar{R} = 2.5$ nm). The highest peak is the scattered pump light in each case. The inset shows a principal sketch of the lowest one-pair transitions involved in the linear absorption. From Woggon et al. (1996b)

in the higher energy range, the PLE signal is composed of all s-, p-, and d-type transitions and the information about the symmetry of the transitions is missing. In this energy part of the spectrum the PP signal has a higher degree of information. The disappearance of the $\Delta\alpha d$ signal in PP around an energy of 2.5 eV, while the PLE signal at that energy is clearly not vanishing, is a clear hint to the p-type character of that transition.

In Figs. 3.16–3.18, the PL, PP and PLE spectra are shown with the pump/detection energy tuned through the linear absorption spectrum of CdSe quantum dots (Woggon et al. 1996b). The results reflect the underlying size distribution in the spectra. In dependence on the pump/detection energy, quantum dots of different sizes were investigated and their properties displayed. The strength of the peaks is proportional to the number of excited crystallites and the oscillator strength of the single transitions.

In Fig. 3.16 the photoluminescence is shown. The inset shows a first-principle calculation of the linear absorption for which at least the three lowest pair states are necessary to compose the spectrum. The line positions have been approximated from the calculation of Fig. 3.11. Pumping the luminescence at $E = 2.175$ eV in the absorption tail, the quantum dots are excited resonantly into their lowest one-pair states. The maximum of the luminescence is red-shifted compared to the excitation. The red shift has been discussed both in terms of a Stokes shift and a sign for a state splitting into one optically allowed (strongly absorbing) state and one optically forbidden (weakly emitting) state (Nirmal et al. 1995; Norris et al. 1995; Chamarro et al. 1995, 1996). Further tuning the pump to higher energies, we come to the situation where two different sizes are involved in the luminescence spectra, one showing the luminescence after excitation in the $(2S_{3/2}, 1s_e)$ state and energy relaxation into the lowest state $(1S_{3/2}, 1s_e)$, the second showing the luminescence from the shifted lowest transition $(1S_{3/2}, 1s_e)$. The energy difference between the two peaks in luminescence represents a rough estimate for the energy difference between the two lowest states. However, due to the size dependence of the red shift, there is no clear assignment possible to a certain size of crystallites. Tuning the pump to even higher energy, a third size of crystallites starts to contribute to the luminescence. The spectrum with the pump at 2.425 eV is composed of signals from very small quantum dots where we resonantly excite the lowest state $(1S_{3/2}, 1s_e)$ and detect the red-shifted luminescence [peak (1)], from quantum dots of larger sizes where we excite into the next excited transition with the hole populating the $2S_{3/2}$-state and detect the luminescence after energy relaxation into the $1S_{3/2}$ hole state [peak (2)], and from quantum dots of very large size, where we excite in any higher state and detect the luminescence after energy relaxation again from the $(1S_{3/2}, 1s_e)$ state [peak (3)]. Peak (3) keeps its position in the spectra even if we tune the energy to still higher values, indicating that the excited states have a very large homogeneous broadening. Peak (1) increases its energy separation to the exciting pump when the dot size decreases. Summarizing, the luminescence qualitatively displays the energy separation between the lowest three excited states, but the lines are very broad and the exact line positions are covered by the size-dependent energy shift between absorption and luminescence and, as we will see below, by the LO-phonon contributions.

Now we turn to the corresponding experiment using nonlinear optics. Figure 3.17 shows the differential absorption $-\Delta\alpha d$ of CdSe quantum dots with an average size of $\bar{R} = 2.3$ nm, i.e. of smaller size than the sample seen in Fig. 3.16. Tuning the pump into the low-energy tail of the linear absorption, we observe a three-band bleaching signal with maxima around 2.24 eV, 2.4 eV and 2.65 eV divided by an induced absorption signal around 2.48 eV. The very small bleaching signal below the pump energy indicates that we are predominantly investigating one size of crystallites. There is no contribution to the bleaching spectrum from dot sizes where the excited states coincide with the pump energy. Tuning the exciting pump laser into the maximum of the linear absorption spectrum, bleaching has been found both at energies below and above the pump energy. In this case again two sizes of quantum dots have been excited, one resonantly into its ground state (the

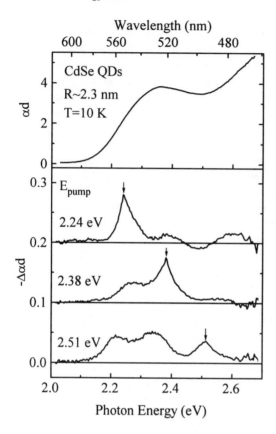

Fig. 3.17. Linear absorption and differential absorption of CdSe quantum dots ($\bar{R} = 2.3$ nm, excitation 2 kW/cm^2). From Woggon et al. (1996b)

bleaching appears on the high-energy side) and the other resonantly into the excited state (the bleaching appears on the low-energy side). Here the induced absorption feature is nearly masked by the bleaching signal. Comparing the energy separation between the first and second hole in differential absorption for the two pump energies at 2.24 and 2.38 eV, similar energy values have been found and thus it can be excluded that the induced absorption had significantly influenced the bleaching maximum of the second hole. Finally, tuning the pump at the energy of 2.51 eV, no bleaching could be observed on the high-energy side, i.e. we are now at the upper end of the size distribution or the excited states are so far in their energy, that they do not contribute any longer to the $\Delta\alpha d$ spectrum. In these measurements the energies of the lowest one-pair transitions can be determined with satisfying exactness. The PP method is well suited to determine the energy states of quantum dots embedded in glasses which do not show efficient luminescence, as well as for small dot sizes with well-separated energy states.

Figure 3.18 shows the differential absorption $-\Delta\alpha d$ of CdSe quantum dots with different sizes. For small quantum dots the probe beam tests only the absorption change of the lowest one-pair transition. For a dot size of

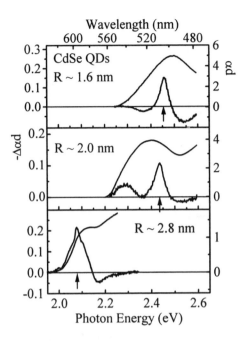

Fig. 3.18. Linear and differential absorption spectra of CdSe quantum dots with different sizes measured at $T = 10\,\mathrm{K}$ and an excitation density of $100\ \mathrm{kW/cm^2}$. The pump energy is indicated by arrows. From Woggon (1996a)

$R = 2\,\mathrm{nm}$ the different hole states are well separated. A well-resolved energetic difference of the two lowest electron–hole-pair transitions of $160\,\mathrm{meV}$ (excitation pump at $2.44\,\mathrm{eV}$) can be observed for the $R = 2\,\mathrm{nm}$ sample. If the excitation is shifted to lower energies (larger dot sizes) this difference decreases (Woggon et al. 1993b). For the sample with CdSe quantum dots of average radii $\bar{R} \sim 2.8\,\mathrm{nm}$, the various 1EHP transitions merge and the bleaching signal becomes broader. In femtosecond hole-burning experiments for CdSe quantum dots of $2.6\,\mathrm{nm}$ radius, an energy distance of the lowest states of $120\,\mathrm{meV}$ could be resolved by Peyghambarian et al. (1989), in spite of the higher excitation intensity and larger laser pulse band-width. For CdS and CdSe quantum dots embedded in glass the published hole burning data mostly show broad bands. Significant influence of the different growth procedures and matrix materials can be seen, for example, in the results of Gaponenko et al. (1993), Peyghambarian et al. (1989), Roussignol et al. (1989), Uhrig et al. (1991), Henneberger et al. (1992), and Woggon and Gaponenko (1995b). Spectrally narrow nonlinear resonances have been only reported for CdSe quantum dots in organic host materials with interfaces well defined by organic groups (Norris et al. 1994; Bawendi et al. 1990b).

Typical PLE spectra are shown in Fig. 3.19. The energy states of the quantum dots with different sizes have been investigated by tuning the detection wavelength through the absorption tail. The PLE spectra are characterized by the transitions to the two lowest pair states $(1S_{3/2}, 1s_e)$ (resonant to the detection energy) and $(2S_{3/2}, 1s_e)$ (the second broad shoulder at higher ener-

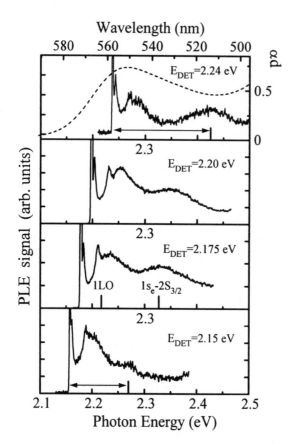

Fig. 3.19. Photoluminescence excitation spectra of CdSe quantum dots ($\bar{R}$ = 2.5 nm) recorded within the tail of the linear absorption spectrum. From Woggon et al. (1996c)

gies). The lowest pair state additionally shows fine structure and LO-phonon replica. When comparing the energy separation between the $(1S_{3/2}, 1s_e)$ and $(2S_{3/2}, 1s_e)$ states for the spectra recorded at 2.24 eV and 2.15 eV the size dependence becomes evident. Furthermore, a size dependence has been detected for the fine structure just near the detection energy. The energy distance of this narrow peak with respect to the detection increases with decreasing size of the quantum dots.

In high-quality CdSe quantum dots produced in organic environment the higher excited states above 2.4 eV have been examined by means of PLE by Norris et al. (1994, 1995). In these high-quality CdSe quantum dots the size dependence of the pair states has been studied both in PLE and PP spectroscopy. Optical transitions have been resolved and assigned to the different excited-hole subbands. Their energy shifts have been followed over a range of dot diameters of 1.9 nm to 11.5 nm. These results likewise confirm that a simple parabolic band approximation cannot account for the observed energy level spacing. Moreover, the avoided level crossing has been observed around

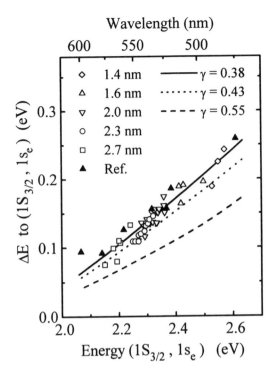

Fig. 3.20. Experimentally determined data for the energy separation between the lowest optical transitions, determined for samples from different sources and by use of the different experimental methods described. The average radii of the various samples are indicated with the corresponding symbols. Additionally, theoretical curves have been plotted as in Fig. 3.11 (for the parameter see text). The filled triangles (Ref.) are data taken from Norris et al. (1994). From Woggon et al. (1996b)

quantum dot sizes of $R = 3.3\,\text{nm}$ intimately connected with the mixture of L and $L + 2$ wave functions (Norris et al. 1994, 1995).

All the experiments confirm that the second excited state for CdSe quantum dots with radii below $0.5a_B$ is the state with the hole populating the $2S_{3/2}$ state. Its oscillator strength is quite strong and can reach about half of the value of the state $(1S_{3/2}, 1s_e)$. The next higher excited one-pair transitions with s-type symmetry are very weak. The $(1P_{3/2}, 1p_e)$ state seems to be much lower in energy compared to the theoretical calculation and dominates the optical properties at energies above the $(2S_{3/2}, 1s_e)$ state in some of the experiments.

In the following we concentrate on the most interesting first excited one-pair state $(2S_{3/2}, 1s_e)$. For comparison with theoretical data, as plotted in Fig. 3.11, the reliability of the experimentally obtained values of $\Delta E = (2S_{3/2}, 1s_e) - (1S_{3/2}, 1s_e)$ can be ensured and the experimental error decreased by deriving the data from the comparison of the PL, PP and PLE experiments (i) to eliminate the two-pair contributions, and (ii) to assign the states to s- and p-type transitions.

In Fig. 3.20 the data obtained from PL, PP and PLE experiments are plotted. The experiments were carried out on a large number of samples of CdSe quantum dots in borosilicate glass matrix where the percentages of the network formers SiO_2 and B_2O_3 and also the amount of the network

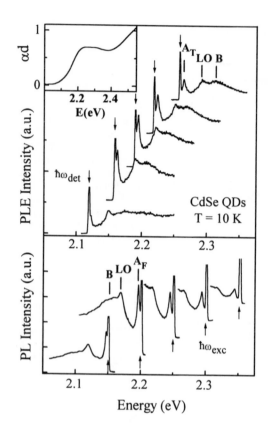

Fig. 3.21. PL and PLE spectra of CdSe quantum dots with $\bar{R} = 2\,\mathrm{nm}$ illustrating the fine structure. From Woggon et al. (1996c)

modifiers K_2O and ZnO have been varied (Müller 1992). This figure can be considered as a summary of present experimental data concerning the energy states of CdSe quantum dots. The axes are chosen according to Fig. 3.11. From Fig. 3.20 it can be concluded that a systematic size-dependence exists for the lowest electron–hole pair transition *independent* of the manufacture and matrix composition, despite the inherent large fluctuations of the experimental values. To support this result, the data obtained by the described experiments for CdSe quantum dots in glass matrix have been plotted along with data from the literature for CdSe quantum dots in organic matrices (data from Norris et al. 1994 in Fig. 3.20), and good agreement can be seen.

The lines in Fig. 3.20 are theoretical curves calculated according to Ekimov et al. (1993). However, the parameter for the valence band coupling γ has been varied (see Fig. 3.11). The best coincidence with the experimental data has been achieved by choosing $\gamma = 0.38$. The model reflects the tendency of the experimental size dependence. Deviations could originate from the use of infinite barriers for the holes, cubic crystal symmetry, and ideal spherical shape as well as from neglecting dielectric effects. For the discussion of the general behavior of the lowest pair transitions, the existence of

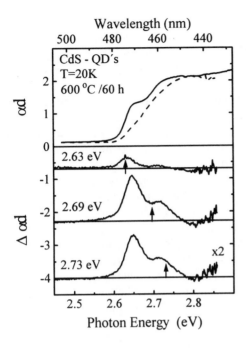

Fig. 3.22. Linear absorption (solid), absorption with the pump of $100\,\mathrm{kW/cm^2}$ at 2.73 eV (dashed), and differential absorption spectra of CdS quantum dots of $R = 7.5\,\mathrm{nm}$ measured at $T = 20\,\mathrm{K}$. The pump energy is indicated by arrows. From Woggon et al. (1993b)

the fine structure is an inherent source of uncertainty. Since in most of the experiments this fine structure is masked by the sample quality, the energy positions have an unavoidable average deviation in the order of this splitting energy.

The effect of fine structure which is presently ascribed to exchange interaction, is demonstrated in Fig. 3.21. The exchange interaction splits the lowest degenerate $(1S_{3/2}, 1s_e)$ pair state into a set of sublevels with the lowest transition being a forbidden one (Norris et al. 1995; Nirmal et al. 1995; Chamarro et al. 1995, 1996, Woggon et al. 1996c). For that reason the states into which absorption and out of which emission occur are separated by an energy gap given by the splitting energy. Figure 3.21 shows the size-dependence of the splitting energy in photoluminescence and photoluminescence excitation measurements with high spectral resolution. Presently, the explanation of this fine structure is the subject of increasing interest, both for II–VI, III–V, and group-IV quantum dots.

Finally, we will give some remarks regarding larger dot sizes in the range $R \geq a_B$. Here, the electronic states are even more difficult to characterize. Luminescence from the ground state becomes optically forbidden in wurtzite-type crystals for certain polarizations of light. Pump-probe spectra show only broad bands caused by the superposition of the whole ladder of excited hole states. The effect of confinement induced valence-band mixing decreases and the lowest energy states correspond to the onset of the confined-level series arising from the top of the 3/2 and the spin–orbit splitoff valence band.

For CdS$_x$Se$_{1-x}$-dot samples with $R \geq a_B$ Uhrig et al. (1992) found that the energy distance of the two lowest main structures in differential absorption spectra approaches the value of the bulk spin–orbit splitting energy Δ_{SO}. For CdSe and CdS, Δ_{SO} is known to be 423 meV and 68 meV, respectively (Landolt–Börnstein 1982).

An example of typical hole-burning spectra obtained for quantum dots in the weak confinement range with $R \geq a_B$ is shown in Fig. 3.22 for CdS-quantum dots with radii of $R = 7.5 \pm 1.5$ nm ($R \approx 2.6\ a_B$, $a_B = 2.8$ nm) (Woggon et al. 1994c, 1995a). The linear absorption spectrum is characterized by two shoulders which are difficult to assign to specific electronic states. However, the differential absorption spectra are dominated by two distinct peaks. Compared to the narrow line-width in the nonlinear absorption spectra of CdSe, the width of the burnt holes is considerably larger in the case of weak confinement due to the excitation into adjacent resonances. Using a fitting procedure based on the superposition of multiple Lorentzian curves, the first absorption maximum has been found at 2.64 eV and related to the electron–hole pair ground state. A second transition can be resolved in the $\Delta\alpha d$ spectra at an energy position of 2.71 eV. The energy distance between the first and the second peak is 70 meV, similar to the valence-band splitting energy Δ_{SO} of bulk CdS due to spin–orbit interaction.

3.2 Two-Electron–Hole-Pair States

3.2.1 Theory

When exciting more than one electron–hole pair inside the quantum dots we necessarily have to deal with electron–hole, hole–hole, and electron–electron interactions, within at least a four-particle system. The number of possible optically excited pairs depends on the density of states available for population, which in turn is determined by the size and confinement range. In big quantum dots with a high number of energy states many pairs can be created leading to the description in terms of a many-body problem. This case will be discussed in the next section. Here small quantum dots will be considered where the spectra have basically only a few well-resolved discrete states. The corresponding Hamilton operator is known as

$$\hat{H} = \hat{H}_e + \hat{H}_h + V_{ee} + V_{hh} + V_{eh} + \delta V(\epsilon_1, \epsilon_2, r_e, r_h) + V_{e,h}^{conf}, \quad (3.31)$$

where $\hat{H}_e$ and $\hat{H}_h$ are the kinetic energies of electrons and holes, V_{ee}, V_{hh} and V_{eh} the Coulomb interaction terms describing the electron–electron, hole–hole, and electron–hole interaction with the background dielectric constant of the bulk semiconductor material, $\delta V(\epsilon_1, \epsilon_2, r_e, r_h)$ is the correction to the Coulomb potential resulting from the differences in the dielectric constants of the semiconductor and the host materials, and $V_{e,h}^{conf}$ is the barrier potential (Hu et al. 1990b; Park et al. 1990).

Two-electron–hole-pair states are stable if their binding energy is positive. The binding energy δE_2 is defined by

$$\delta E_2 = 2E_1 - E_2 , \qquad (3.32)$$

with E_1, E_2 being the one- and two-pair ground-state energy, respectively. The two-pair ground state is characterized by antiparallel oriented spins for electron and holes. When we omit the Coulomb terms and considering only the kinetic energies $\hat{H}_e + \hat{H}_h$, the one- and two-pair energy states are degenerate and $\delta E_2 = 0$. Bound two-pair states can be obtained only by including the Coulomb potentials.

In bulk semiconductors the states obtained by involving the Coulomb interaction were denoted as exciton and biexciton, respectively. These terms have often been used also for quantum dots to describe the one-electron–hole-pair and two-electron–hole-pair states and imply that the corresponding energy states should evolve into the bulk exciton and biexciton energies with increasing dot size. In quantum dots, however, the energy $2E_1$ is not a correct eigenenergy because the biexciton cannot dissociate into free electrons and holes because of the confining barriers. Thus (3.32) is only a first approach.

To solve the Schrödinger equation derived from (3.31), the numerical matrix diagonalization has been rendered again as an efficient method (Hu et al. 1990a,b; Park et al. 1990). Beside the matrix diagonalization also perturbation theory, variational calculations and Monte-Carlo technique have been applied to describe the two-electron–hole-pair states (Banyai et al. 1988a,b; Banyai 1989; Takagahara 1989; Efros and Rodina 1989; Pollock and Koch 1991).

In the intermediate case of confinement and for $m_e/m_h < 1$ so that $a_h < R < a_e$ holds, the potential acting on the holes can be approximated by an effective Coulomb potential from the averaged field of the confined electron states (Efros and Efros 1982). A numerical calculation by Banyai et al. (1988a) yields for the biexciton energy within this model

$$\delta E_2 \simeq \left[0.4 - 1.8 \sqrt{\frac{a_h}{R}} \right] \frac{a_B}{R} E_{Ry} , \qquad (3.33)$$

a strictly positive binding energy δE_2. Likewise, detailed numerical calculations made by Takagahara (1989), Pollock and Koch (1991), and (Hu et al. 1990a,b) showed that two-pair states are stable in quantum dots. The value of the binding energy depends sensitively on the differences in the dielectric constants of the host and the matrix material. The smaller value of ϵ_1 outside the semiconductor dot weakens the screening of the Coulomb interaction between electrons and holes and thus enhances their binding energy. Compared with the 3D case, the theoretical calculations predict for biexcitons in three-dimensionally confined quantum dots (i) a larger binding energy, (ii) an observation of both ground- and excited two-pair states in the optical spectra because of the absence of the 3D- typical continuum states and (iii) transitions to excited two-pair states which were originally forbidden and occur

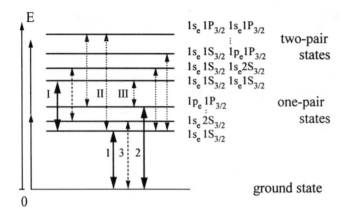

Fig. 3.23. Optical transitions for a two-particle system in a quantum dot

now due to the change of selection rules caused by the Coulomb potential. Therefore, the two-electron–hole-pair or biexciton states are an important elementary excitation, especially in small dots.

When we consider in addition the transitions between one-electron–hole-pair states (or excitons) and two-electron–hole-pair states (or biexcitons), the scheme of optical transitions changes according to Fig. 3.23. Because of the occurrence of a whole ensemble of one-pair states, the two-pair states also comprise a multitude of states and transitions between them. Beside the combination $(1S_{3/2}, 1s_e, 1S_{3/2}, 1s_e)$ for the lowest pair state, combinations also appear with the hole or the electron populating excited states, for example, $(1S_{3/2}, 1s_e, 2S_{3/2}, 1s_e)$.

The strongest two-pair transition (I) appears energetically below the lowest one-pair resonance (1), separated by the biexciton binding energy. The weaker transitions involving excited electron and hole states can be observed both low-energetically (III) and high-energetically (II) with respect to the lowest one-pair transition (compare with Figs. 3.3 and 3.7). The evidence of stable biexciton states can be confirmed by investigating the nonlinear absorption spectra. In Fig. 3.24, the change in absorption $-\Delta\alpha$ has been calculated for a quantum dot of $R = a_B$ (Hu et al. 1990b). The other parameters are $\epsilon_2/\epsilon_1 = 1$, and $m_e/m_h = 0.24$. The result is shown in the energy range around the lowest one-pair resonance for different damping constants γ and a pump-beam resonant to the one-pair energy. The positive change in $-\Delta\alpha$ indicates the saturation of the one-pair resonance. The induced absorption on the high-energy side of the one-pair bleaching is caused by transitions to excited two-pair states where one or both holes are not in their ground state. These transitions become possible since the Coulomb interaction changes the selection rules. The generation of two electron–hole pairs is proceeded by an

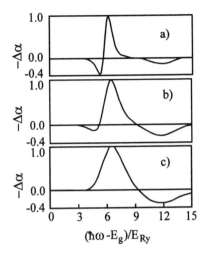

Fig. 3.24. Differential absorption $-\Delta\alpha$ showing the saturation of the one-pair-states and the induced absorption of the two-pair states calculated with $R/a_B = 1$, $\epsilon_2/\epsilon_1 = 1$, $m_e/m_h = 0.24$ and different damping constants (a) $\gamma = E_{Ry}$, (b) $\gamma = 2E_{Ry}$ and (c) $\gamma = 3E_{Ry}$, respectively. From Hu et al. (1990b)

absorption process involving one pump and one probe photon. The induced absorption on the low-energy side is assigned to the biexciton ground state.

Before discussing experimental results, the effect of $\delta V(\epsilon_1, \epsilon_2, r_e, r_h)$ on the biexciton binding energy should be mentioned. The additional Coulomb forces due to the surface polarization help to stabilize the two-pair states. Calculations have been carried out by Hu et al. (1990b) varying the ratio in the dielectric constants between $\epsilon_2/\epsilon_1 = 1$ and $\epsilon_2/\epsilon_1 = 10$. The induced absorption feature becomes more and more pronounced with increasing ratio ϵ_2/ϵ_1 (ϵ_2 is the dielectric constant of the semiconductor dot and ϵ_1 of the host).

3.2.2 Experiments to Identify Two-Pair States

First hints to the occurrence of excited two-pair states have been obtained from luminescence spectra of $CdS_{1-x}Se_x$ quantum dots under high laser excitation (Uhrig et al. 1990). Structures in the high energy wing were interpreted as transitions arising from the decay of excited two-electron–hole-pair states.

By means of nonlinear absorption measurements the induced absorption peak above the one-pair resonance was observed by several authors confirming the existence of transitions with the excited two-pair states as final states which are not dipole-allowed in the bulk material (Park et al. 1990; Peyghambarian et al. 1989; Woggon et al. 1993b). The expected induced absorption related to the biexciton ground-state *below* the one-pair resonance, which would confirm the positive binding energy, is difficult to observe. The resonance is visible only for very small homogeneous broadening parameters γ (see Fig. 3.24), and for quantum dot sizes where the different hole states are well separated in energy. Saturating one-pair resonances tend to cover the induced absorption feature on the low-energy side.

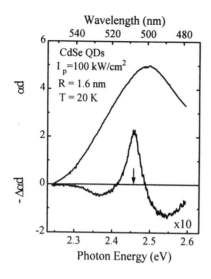

Fig. 3.25. Linear and differential absorption spectra $-\Delta\alpha d$ of CdSe quantum dots of $R = 1.6$ nm measured at $T = 20$ K and an excitation density of $100\,\text{kW/cm}^2$. The pump energy is indicated by an arrow. Data taken from Woggon et al. (1994a, 1995a)

Figure 3.25 shows the nonlinear change in the absorption $-\Delta\alpha d$ for CdSe quantum dots with high-quality surfaces and with $R = 1.6$ nm characterized by very small values γ of the homogeneous line-width (Woggon 1994b, 1995a). For these small sizes the energetic distance between the lowest one-electron–hole-pair states is larger than 200 meV and therefore only the transition involving the first 1EHP state $(1S_{3/2}, 1s_e)$ can be seen in the differential absorption spectrum. For a laser excitation intensity of $100\,\text{kW/cm}^2$, spectrally narrow nonlinear resonances in the differential absorption have been found that shift with excitation energy within the spectral width of the linear absorption peak. The spectral shape of the differential absorption agrees very well with that calculated by Hu et al. (1990a,b) (Fig. 3.24). The narrow *bleaching* corresponds to the saturation of the one-electron–hole-pair transition of the resonantly excited quantum dots, while the *induced* absorptions on the low- and high-energy side are attributed to the creation of two-pair states from the one-pair states.

Another possible way to observe the ground state biexciton is the use of an additional saturator-beam besides the pump and probe beams in a three-beam experiment (Kang et al. 1993). The saturator beam has been tuned below the pump energy to bleach the one-pair transition of quantum dots of certain size. Because the signal from these quantum dots does not contribute to the nonlinear absorption spectrum, the induced absorption of the two-pair ground state can then be resolved. Biexciton effects in femtosecond pump-and-probe spectra have been reported by Klimov et al. (1994) for CdSe quantum dots.

The bulk semiconductor binding energy of biexcitons in II–VI compounds is very small [for CdS holds $\delta E_2 = 5.7$ meV; for CdSe holds $\delta E_2 = 1.2...5$ meV

(Landolt–Börnstein 1982)]. Even assuming an increase of the biexciton binding energy with confinement of more than twice the bulk excitonic Rydberg energy, the effect in the spectra is less pronounced than, for example, expected for materials with higher bulk biexciton binding energies. Promising candidates for the experimental proof of the increase in δE_2 are therefore I–VII semiconductors like CuCl ($\delta E_2 = 28\,\text{meV}$) or CuBr ($\delta E_2 = 19.5\,\text{meV}$). The drawback is the small corresponding excitonic Bohr radius which already yields for the one-pair states only a weak confinement effect. The quantum confinement results in a large number of states in a narrow interval of energy. Using the pump-probe experimental technique, the different states can hardly be distinguished. Furthermore, the two-photon absorption (TPA), well-suited for detection of biexcitons in bulk CuCl and CuBr (Nagasawa et al. 1976; Nozue et al. 1982), has failed up to now in CuBr and CuCl quantum dots. This is probably caused by the high number of additional states resonant to the biexciton energy, which are produced by traps or by dots of other sizes, as well as by the low filling factor of the quantum dots in the matrix.

Compared to these experiments, the investigation of luminescence and the analysis of the intensity dependence via kinetic models remain the most valuable experimental methods for the investigation of biexcitons. We will outline here the results obtained for CuBr quantum dots.

The intensive work done for bulk CuBr to characterize the excitonic polariton and the biexciton resulted in a detailed knowledge of the energy values in the limit of large quantum dots (for a review see for example Nagasawa et al. 1976, 1989; Nozue et al. 1982; Hönerlage et al. 1980, 1985; Grun et al. 1976). As in the case of most semiconductors, CuBr crystallizes in zinc-blende structure with the lowest conduction band of Γ_6 symmetry and the two highest valence bands of Γ_8 and Γ_7 symmetry. Excitons arising from the upper two valence bands are often denoted as $Z_{1,2}$ and Z_3 excitons, respectively. In CuBr the degeneracy of the exciton states is lifted because of the strong exchange interaction between electron and hole, resulting in the corresponding lowest exciton energies $E(\Gamma_{5L}) = 2.9766\,\text{eV}$, $E(\Gamma_{5T}) = 2.9644\,\text{eV}$ and $E(\Gamma_{3,4}) = 2.9627\,\text{eV}$ (see Fig. 3.26). The exciton binding energy $E_{\text{Ry}}^{\text{ex}}$ and the Bohr radius a_B in CuBr are $108\,\text{meV}$ and $1.25\,\text{nm}$, respectively.

The lifting of degeneracy for the exciton states is preserved also for the biexciton states. Thus, the biexciton–exciton transition possesses different possible energies. For CuBr, a M_T and a M_L band have been found in luminescence attributed to the decay of the biexciton ground state into the transverse and longitudinal exciton, respectively, and also a M_f-band related to the biexciton–exciton transition where the final state is the $\Gamma_{3,4}$-exciton. In its ground state of Γ_1 symmetry, the biexciton has a binding energy of $19.5\,\text{meV}$ with respect to the continuum of two excitons of Γ_3,Γ_4 symmetry (Hönerlage et al. 1985). The splitting of the biexciton ground state into Γ_1, Γ_5 and Γ_3 states, the dispersion and the behavior of biexcitons in magnetic

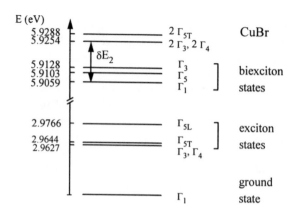

Fig. 3.26. The exciton and biexciton energy levels in bulk CuBr. Data taken from Hönerlage et al. (1985)

fields (Nagasawa et al. 1976; Hönerlage et al. 1985) have been revealed by the analysis of TPA, and a summary of the data is given in Fig. 3.26.

Because of the small size of the quantum dots, the polariton-like properties are usually neglected and the exciton and biexciton states in quantum dots treated within the framework of weak exciton–photon coupling as discussed by Hu et al. (1990b), Banyai (1989), Takagahara (1989), and Efros and Rodina (1989). Additionally, size variations prevent the observation of splitting effects of the order of magnitude of only a few meV.

We thus temporarily neglect the splitting of the 3D exciton and biexciton levels. As the relevant transitions in CuBr quantum dots we consider for excitons the transition from the Γ_3, Γ_4, (Γ_{5T}) exciton state ($E \approx 2.964\,\text{eV}$) to the Γ_1 ground state and for the biexcitons the decay from the Γ_1 ground state ($E \approx 5.906\,\text{eV}$) into an exciton of Γ_3, Γ_4, (Γ_{5T}) symmetry and a photon. In the following we investigate the steady-state and time-resolved luminescence of CuBr quantum dots from different growth processes and in the wide range of sizes between $R/a_B \approx 1$ to 6 nm. The growth parameters are summarized in Table 3.2. The exciton-related transitions in luminescence were identified by comparison with the corresponding absorption peaks of the CuBr quantum dots investigated. In Fig. 3.27, the absorption spectra are shown for samples with different radii, namely $R = 6.75$, 3.8, and 1.8 nm. The position of the energetically highest peak in the luminescence of each sample is indicated by arrows.

The comparison of the maxima of the absorption and of the luminescence for the size range investigated leads to a classification of the samples into two groups depending on size. When we decrease the radius from 6.75 nm to $R \approx 3$ nm, the shift of the highest luminescence peak strictly follows the confinement-induced shift of the $Z_{1,2}$-absorption maximum. For these samples in the range of sizes between 6.75 nm to 3 nm, the energetically highest luminescence band has always been found to be around 20 meV below the absorption peak (see Fig. 3.27, curve 1 and 2). However, for a further decrease in size the shift in luminescence converges to the spectral behavior

Table 3.2. Parameter of the investigated CuBr doped glasses and energetic distance of the luminescence of the H and L peak for the two growth series of the CuBr quantum dots; data taken from Woggon et al. (1994d)

Heat treatment	R/nm	R/a_B	E_H/eV	E_L/eV	Δ/meV
Series 1					
550 °C/ 1h	1.4	1.12	3.042	2.985	57
550 °C/ 3.2h	2.0	1.6	3.032	2.981	51
550 °C/ 8h	3.0	2.4	3.018	2.976	42
550 °C/ 27h	4.4	3.52	3.002	2.968	34
550 °C/ 43h	5.2	4.16	2.989	2.965	24
Series 2					
475 °C/ 1h	1.8	1.44	3.038	2.988	50
540 °C/ 1h	3.8	3.04	3.006	2.970	36
580 °C/ 1h	6.75	5.4	2.975	2.950	25

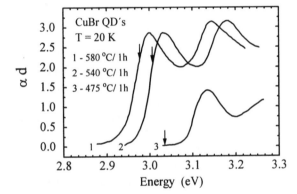

Fig. 3.27. Absorption spectra of CuBr quantum dots of different radii: (1) 6.75 nm, (2) 3.8 nm; (3) 1.8 nm. The energy of the corresponding luminescence peak is indicated by arrows. From Woggon et al. (1994d)

similar to the luminescence obtained from the glass matrix before annealing, whereas the absorption peak shifts further to higher energies as the result of the confinement (see curve 3 in Fig. 3.27). Obviously, for this second group of samples with sizes $R \leq 3$ nm in which a considerable number of atoms are situated at the interface, there is a high density of initial and final states for optical transitions not influenced by size variation and confinement.

For all samples of Table 3.2, quasi-steady state luminescence measurements have been performed using ns-laser pulses ($\lambda_{exc} = 337$ nm) with 45° angle of incidence for excitation and detection of the emission from the front side to avoid reabsorption. Figure 3.28 shows the typical luminescence spectra at different excitation intensities for a selected sample of $R = 3$ nm. The spectra consist of two pronounced luminescence peaks (in the following labelled H peak for the higher energetic one and L peak for the low-energetic peak), in particular in samples of series 1, but also seen in the samples of series 2.

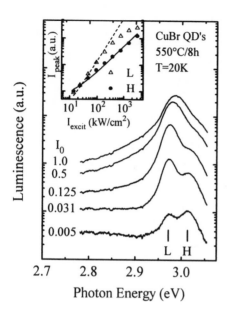

Fig. 3.28. Intensity dependent luminescence spectra of CuBr quantum dots of $R = 3\,\mathrm{nm}$, $T = 20$ K. The insert shows the dependence of the luminescence intensity on the excitation intensity for the biexciton peak (triangles, L peak) and the exciton peak (circles, H peak). From Woggon et al. (1994d)

To identify the two peaks H and L as the exciton and biexciton states, the intensity and time-dependence of the luminescence spectra and the change in energy with sizes have been investigated. For the highest intensity investigated the shift of the whole spectrum to higher energies indicates the onset of nonlinear optical effects in the corresponding absorption spectrum resulting in a blue shift of the $Z_{1,2}$ absorption peak in CuBr (Woggon et al. 1988). To avoid this influence, we compare the luminescence spectra at an excitation intensity around $250\,\mathrm{kW/cm^2}$ in the following discussion.

Using two Lorentzians, a good fit can be obtained for the luminescence spectra measured at this intensity and the existence of two different peaks has been confirmed by a careful line shape analysis. The corresponding energies of the maxima E_L and E_H are given in Table 3.2. For large dot sizes, the energy of the L peak matches the bulk value of the biexciton luminescence as the H peak does with regard to the exciton luminescence. A further hint at the biexcitonic nature of the luminescence of the L peak is the superlinear dependence of the luminescence signal on the excitation intensity shown in the insert in Fig. 3.28 (the ideal exciton–biexciton transition should result in a I^2 dependence of the biexciton luminescence peak).

Concerning the influence of the different growth processes, two remarks can be made about the results obtained: (i) the two peaks H and L appear independently from the growth process in all samples, and (ii) whereas the absorption spectra show no significant influence on the modification of the growth process, the different melting and annealing procedures result for samples of series 2 in an additional broad luminescence band shifted to low energies with a tail partly covering the H and L peaks. The luminescence

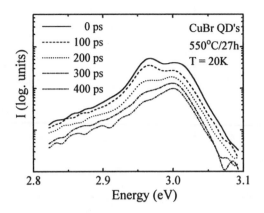

Fig. 3.29. Spectral changes of the luminescence at different delay times. From Woggon et al. (1994d)

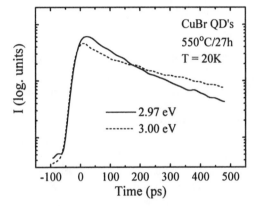

Fig. 3.30. Decay curves measured at the energy of the exciton and biexciton peak. From Woggon et al. (1994d)

efficiency of the H and L peaks is very low in these samples and decreases rapidly under illumination. These differences have been attributed to various trap states caused by the growth process.

For the time-resolved luminescence of the exciton–biexciton system, one expects an instantaneous rise of the excitonic luminescence followed by the decay of one part of the excitons into the ground state. The other part forms biexcitons accompanied by the detection of biexcitonic luminescence in the spectra. For the luminescence of the biexciton system a slower rise time is expected because of the finite formation time for biexcitons. At moderate excitation intensity, this scenario leads to the coexistence of excitons and biexcitons. If the decay of the luminescence is only determined by the quantum mechanical transition probabilities of the excitons and biexcitons, then the biexciton lifetime should be shorter than the exciton lifetime. At very high excitation intensities, density-dependent recombination channels will be

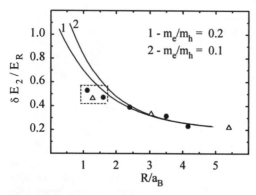

Fig. 3.31. Comparison of the biexciton binding energy observed in CuBr quantum dots with the theory of Hu et al. (1990b). The deviation at the smallest dot radii (indicated by the dashed box) is attributed to the breakdown of the confinement concept for the luminescence (see text). Full circles – samples of growth series 1; triangles – samples of growth series 2. From Woggon et al. (1994d)

opened, for example, by collision processes decreasing the carrier lifetime in both the exciton and biexciton systems.

For bulk CuBr the time-resolved luminescence of biexcitons has been measured by Masumoto et al. (1984) after band-to-band excitation. The delayed rise time of the biexciton system as well as the shorter decay compared to excitons is reflected in the experiment. The peak of the biexciton luminescence is delayed about 100 ps with respect to the maximum of the exciton luminescence, and an analysis of the decay curves by rate equations results in lifetimes of 150 ps and 60 ps for excitons and biexcitons in 3D CuBr ($T = 4.2$ K).

The results obtained for CuBr-quantum dots concerning the time-resolved luminescence experiments (characterized in Table 3.2) are very similar to the work on CuBr bulk material of Masumoto et al. (1984). Figure 3.29 shows the time-resolved luminescence at moderate excitation intensities of 1 mJ/cm^2 for the sample annealed at 550 °C/27 h, and Fig. 3.30 the decay curves at the energy positions of the H peak ($E = 3.00$ eV) and the L peak ($E = 2.97$ eV) of the same sample. The temporal behavior reflected in the experiments is in good agreement with the considerations above. The delayed rising process is clearly shown by the temporal shift of the maximum of the L peak in Fig. 3.30. The decay of the L peak is faster than that of the H peak.

The biexciton binding energy for the different dot radii have been determined from the distance Δ of the L and H peak in the luminescence and given in Table 3.2. Figure 3.31 shows the comparison of the experimental results of the luminescence experiments presented here (Table 3.2) and the theory by Hu et al. (1990a,b). An excellent agreement can be seen. The deviation at the smallest radii of quantum dots (indicated by a dashed box) is attributed to the breakdown of the confinement concept because of surface-related trapping effects for the luminescence of the smallest nanocrystals.

The observation of biexciton luminescence for CuCl quantum dots has been reported by Levy et al. (1991), Itoh et al. (1989), Masumoto et al. (1994a,b), and Edamatsu et al. (1995) supporting the theoretical result of increasing biexciton binding energy with confinement. Recently, also a lasing process of biexcitonic origin has been observed in CuCl quantum dots by Masumoto et al. (1993) and Faller et al (1993).

In conclusion, the exciton and biexciton states have been found in the steady-state luminescence. An increasing energetical distance between exciton and biexciton peak has been found in the spectra for decreasing dot radius. This behavior supports the theoretical predicted increase of the two-pair binding energy δE_2. Furthermore, this interpretation is confirmed by the dynamical behavior of the respective emission bands.

3.2.3 Optical Gain

The formation of two-pair states has already been demonstrated by the appearance of an induced absorption around the bleached transition in the differential absorption spectra, as discussed in Fig. 3.15 and Fig. 3.25. It is important to consider this point when searching for the mechanism of optical gain in quantum dots. In this section we analyze the gain spectra obtained for strongly confined CdSe quantum dots. To reveal the physical origin, the gain and bleaching spectra are compared with the corresponding luminescence spectra and with picosecond time-resolved luminescence under very high excitation (Wind et al. 1996). Gain in CdSe nanocrystals with sizes above the Bohr radius was reported some years ago by Dneprovskii et al. (1992a) and Vandyshev et al. (1992). For quantum dots in the strong confinement regime, optical gain could be detected in femtosecond differential absorption (Giessen et al. 1994, 1995).

Figure 3.32 shows gain spectra of CdSe quantum dots in the strong confinement range. The average radius in the selected sample is ~2.7 nm, which is about half the excitonic Bohr radius of 5.8 nm in bulk CdSe. In differential absorption experiments the changes in absorption have been measured when the samples were pumped above the absorption edge at 2.48 eV with ns pulses. For the highest intensities, the signal in differential absorption exceeds the linear absorption and yields the gain spectrum. The gain is spectrally broad (nearly 200 meV) with a steeper decrease on the high energy side and a longer tail stretching to lower energies.

A qualitative scheme of the transitions expected for strongly confined quantum dots is shown in the inset of Fig. 3.32. Besides the transition between the lowest one-pair state $(1S_{3/2}, 1s_e)$ and the quantum dot ground state, a whole ensemble of transitions from higher one-pair states involving excited hole states are allowed due to the relaxation of the selection rules by, for example, the Coulomb interaction and valence-band mixing effects. For the

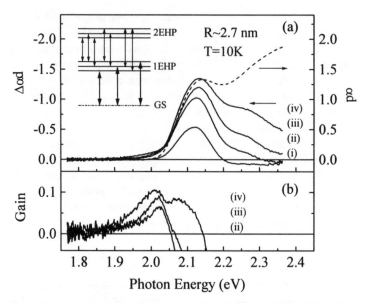

Fig. 3.32. (a) Linear absorption αd (dashed) and differential absorption $\Delta\alpha d$ (solid) for pump intensities of (i) 10, (ii) 100, (iii) 400, and (iv) 1400 kW/cm². The inset shows a scheme of the optical transitions involving one-pair and two-pair states. (b) Gain spectra in units of $\Delta\alpha d$ for pump intensities as in (a). From Wind et al. (1996)

same reason, two-pair states (or biexcitons) can also very likely be formed and consequently transitions between these two-pair states, both from their lowest level and from various excited levels, and the one-pair states will result. These transitions can therefore appear both on the low- and high-energy side of the transitions from one-pair states (or excitons) in the spectrum. The spectral shape of the gain in quantum dots is dramatically changed by this multitude of stimulated transitions arising from the biexciton states. A broad gain spectrum will appear instead of sharp discrete lines as a new typical feature of stimulation in zero-dimensional systems. This qualitative discussion illustrated in Fig. 3.32 has also been treated theoretically by Hu et al. (1996) by a full, numerical approach yielding exactly the very broad gain spectrum.

In Fig. 3.33a, the numerically calculated differential absorption spectra are shown with the excitation intensity as the parameter for a dot size of $R = 0.5a_B$. In the insert of Fig. 3.33a the oscillator strength of the transitions from the 1EHP states to the ground state and from the 2EHP states to the 1EHP state is shown. For clarity the former is plotted above the baseline, the latter below it. As can be seen, the modification in the ideal selection rules owing to the Coulomb interaction and the valence-band mixing results in a multitude of new optical transitions. At high excitation the nonlinear spectra of these transitions overlap and form the differential absorption signal

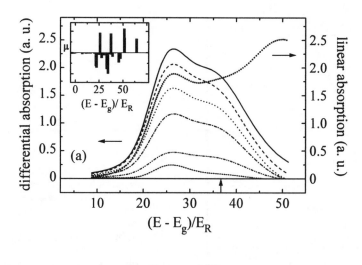

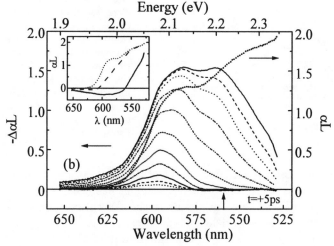

Fig. 3.33. (a) Calculated differential absorption spectra with the pump intensity at 37 $(E - E_g)/E_R$ as the parameter. From top to bottom the curves correspond to decreasing one-pair and two-pair population. The linear absorption (right scale) has been calculated considering a size distribution of 10 %. The insert show the dipole matrix elements μ for the one-pair transitions (above the baseline) and the two-pair transitions (below). (b) Experimental differential absorption for different intensities I_0, $I_0/2$, $I_0/4$, $I_0/8$, ...etc. ($I_0 = 25\,\mathrm{mJ/cm^2}$, $T = 10$ K, pump at 560 nm). The insert shows the linear absorption and the absorption spectra at I_0 (solid) and $I_0/32$ (dashed). From Giessen et al. (1995)

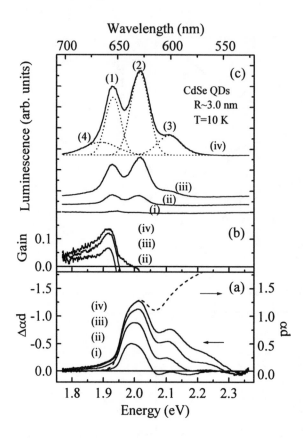

Fig. 3.34. (a) Linear absorption αd (dashed) and differential absorption $\Delta\alpha d$ (solid) for pump intensities of (i) 10, (ii) 100, (iii) 400, and (iv) 1400 kW/cm². (b) Gain spectra in units of $\Delta\alpha d$ for pump intensities as in (a). (c) Evolution of the luminescence spectra (solid) for identical excitation intensities as in (a). The dotted lines show a deconvolution of curve (iv) for the four luminescence peaks. From Wind et al. (1996)

plotted in Fig. 3.33a (Giessen et al. 1995, Hu et al. 1996). In order to verify the calculations, a femtosecond pump-probe experiment has been performed and the result is shown in Fig. 3.33b. The change in the absorption $-\Delta\alpha L$ has been measured with the probe pulse delayed 5 ps after the 115 fs pump pulse and for various excitation intensities. The experimental data agree very well with the calculations. The inset to Fig. 33b shows the same experimental data, except that the probe beam absorption αL is plotted instead of the $-\Delta\alpha L$ signal. In the case of maximum pump intensity I_0 (solid line), the gain has a maximum value of $-\alpha L = 0.28$ (22 % of the linear absorption).

To support this model of a biexcitonic origin of the gain further, in the following the intensity-dependent bleaching of the absorption will be compared with the luminescence spectra obtained under identical experimental conditions.

Figure 3.34a shows, besides the linear absorption spectrum αd, the development of the differential absorption $\Delta\alpha d$ with increasing pump intensity. The pump photon energy of 2.48 eV was chosen well above the onset of the absorption spectrum to excite all quantum dots within their size distribution. To achieve optical gain in quantum dots, it is necessary to bleach the whole size distribution, which is feasible by exciting into the dense-lying higher states with a large homogeneous line-width. Otherwise, with size-selective excitation, the stimulated emission from one size of dots would be reabsorbed by dots of other sizes if their corresponding transitions were not bleached. For the lowest pump intensity of $10\,\mathrm{kW/cm^2}$ (i), the absorption is bleached around $2.0\,\mathrm{eV}$, indicating the relaxation of the excited electron–hole pairs to the lowest state $(1S_{3/2}, 1s_e)$. The other hole levels belonging to the same electron level $1s_e$ are bleached simultaneously.

For higher pump intensities [curves (ii)–(iv)], the bleaching signal is increased up to the same value as the maximum of the linear absorption, showing that the whole size distribution has indeed been excited. At the same time, below the lowest one-pair transition the magnitude of the differential absorption exceeds the linear absorption of the nonexcited sample indicating optical gain in a spectral range where no states existed before. To show the spectral shape of the gain, the difference of the linear absorption and the differential absorption are plotted for different pump intensities in Fig. 3.34b. With increasing intensity the gain increases and becomes spectrally broader. The maximum of the gain for the highest intensity amounts to about $6.5\,\mathrm{cm^{-1}}$, which is more than $10\,\%$ of the absorption peak. The gain spectra are now compared with the luminescence taken under identical excitation conditions shown in Fig. 3.34c. For the lowest intensity (i) only one luminescence peak (1) near the absorption edge is detected. It can be assigned to the radiative recombination of electron–hole pairs after relaxation into the lowest quantum dot level. At higher pump intensities [curves (ii)–(iv)], this peak grows and, at the same time a second peak (2) $\sim 80\,\mathrm{meV}$ above develops, which finally even exceeds the first one originating from the decay of a one-pair state with the hole in the $2S_{3/2}$ level. Additionally, a third peak (3) $\sim 180\,\mathrm{meV}$ emerges above the first one [clearly seen in curve (iv) of Fig. 3.34c]. Regarding the luminescence peak (3) evolving at high pump intensity, contributions to this line can be expected from higher excited two-pair states (biexcitons with holes populating excited states) and from a p–p-type transition.

As the most interesting fact, a shoulder (4) appears on the low-energy side of the first peak (1) in the spectral region where the gain is found. A careful deconvolution of the luminescence spectrum for the highest pump intensity shows that the spectrum can be fitted excellently with a sum over four Gaussians, indicating that the low-energetic shoulder (4) arises from a peak centered about $40\,\mathrm{meV}$ below the first transition. This peak shows a much larger half-width (of $\sim 100\,\mathrm{meV}$) compared to peaks (1) and (2) (of $\sim 50\,\mathrm{meV}$ and $\sim 60\,\mathrm{meV}$).

From the spectral position of its appearance under high excitation, from the coincidence of its energetic position with the gain and from its large width, we conclude that this new luminescence peak (4) arises from the decay of two-pair states (biexcitons) into a photon and an exciton.

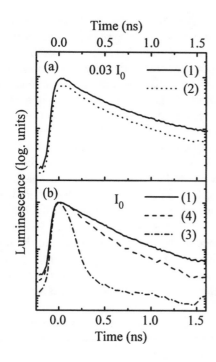

Fig. 3.35. Decay curves of the luminescence measured at the spectral positions of the luminescence peaks marked in Fig. 34(c) for the fluences (a) 0.03 I_0 and (b) I_0. From Wind et al. (1996)

As bulk CdSe has a low biexciton binding energy of ~5 meV (Schwab et al. 1992), the biexciton luminescence merges with the exciton transition for large dot sizes. But for dots with radii of half the excitonic Bohr radius, an enhancement of the biexciton binding energy up to a factor of 5 has been predicted by theory (Hu et al. 1990a,b). Furthermore, the decay of a two-pair state into a photon and a one-pair state can leave the one-pair state not only in its ground level but also in various excited-state levels, emitting then a photon of lower energy. This can explain the breadth of peak (4) in the luminescence as well as the long tail of the gain spectra.

Arguments supporting the assumption of a biexciton decay can also be obtained from time-resolved luminescence. In the time-resolved experiments the luminescence was excited by 70 ps (FWHM) long pulses from a quenched-transient dye laser with the same photon energy of 2.48 eV. For detection, a combination of a spectrometer and a streak camera was used.

Figure 35 shows the temporal decay of the luminescence at the spectral position (1) to (4) from Fig. 34c for two different pump fluences. For a fluence of 0.03 I_0 the excitation is comparable to curve (ii) in Fig. 3.34c showing the two peaks from the $(1S_{3/2}, 1s_e)$ and $(2S_{3/2}, 1s_e)$ states. The decay time of both peaks is nearly identical, as can be seen from the decay curves of Fig. 3.35a. But for the high fluence I_0 of 5 mJ/cm^2, the very fast and nonexponential decay of the high energy peak (3) in the luminescence decay of Fig. 3.35b indicates that the nature of this emission is completely different from that of the two peaks at low fluence. This supports the assumption of its origin either from p–p-type transitions or from excited two-pair states. The new peak (4) appearing on the low-energy side also has a faster decay rate than the $(1S_{3/2}, 1s_e)$ transition (1). This fact is understandable if we assume the formation of biexcitons at high excitation and their decay with emission energies shifted by the biexciton binding energy.

Similar behavior has been described for the biexciton emission in CuBr quantum dots (Woggon et al. 1994d) and CuCl quantum dots (Leonelli et al. 1992; Faller et al. 1993; Masumoto et al. 1993). Thus, the emission spectra under high excitation are characterized by radiative processes connected with the biexciton system and appearing on the high- and low-energy side of the $(1S_{3/2}, 1s_e)$ and $(2S_{3/2}, 1s_e)$ exciton states. These results support the explanation given for the gain in differential absorption.

3.3 Many Particle Interaction

Within the particle-in-an-infinite-box problem, the number of energy states and their energy distance depend on the height of the barrier potential and are proportional to $\sim 1/R^2$ and $\sim 1/m$, the radius of the sphere and the mass of the particle. Increasing the radius of the quantum dot, a rising number of energetically close states appears and very rapidly a value of some hundreds of states per unit of energy can be reached. The number of wave functions which have to be considered for a numerical treatment (e.g. based on matrix diagonalization methods) is very high.

The weak confinement model proposed in the early work of Efros and Efros (1982) has proved to be a good description of the optical properties of large quantum dots. The electron–hole-pair states have been treated assuming that the confinement does not act on the relative coordinates of electron and holes and leads only to a quantized center-of-mass motion. Analyzing the linear absorption, this concept has been successfully applied to large quantum dots (i.e. large with respect to the excitonic Bohr radius) of semiconductor materials with high binding energy of excitons and biexcitons like CuCl and CuBr (Ekimov et al. 1985b,1986; Itoh et al. 1988a,b). In contrast to the bulk, where the Lorentzian line shape is hard to observe since the exciton states are strongly coupled to the field of the electromagnetic wave, the diluted semiconductor system embedded in a glass matrix shows well-resolved excitonic resonances. Typically the $Z_{1,2}$ and Z_3 exciton peaks dominate in the spectra of quantum dots, but with energies slightly shifted to higher values. The lineshape is characterized by an asymmetric high energetic wing resulting from the underlying series of confined overlapping energy levels (see also Fig. 3.27 and Henneberger et al. 1988).

Evidently, under weak confinement conditions we expect that the nonlinear-optical behavior will be determined by the many-particle system and its possible interactions. The two-level model, often useful to describe the

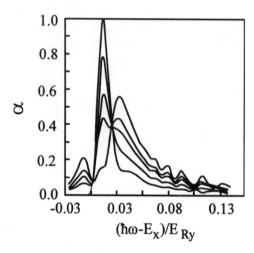

Fig. 3.36. Theoretical absorption spectra calculated for different populations of the one-exciton state ($R = 10$ a_B, appropriate to CuCl) with different weights of the occupation (between 0 and 1), at $T = 10$ K and homogeneous width $\Gamma = 0.005$ E_{Ry}. From Belleguie and Banyai (1991)

nonlinear optics of small quantum dots with well-resolved peak structures, will fail here.

Based on the weak confinement model, Belleguie and Banyai (1991) started the description of the optical nonlinearity from a boson picture of the exciton system in quantum dots of large radius ($R \gg a_B$). In contrast to electrons and holes which are fermions, excitons and biexcitons have integer spin quantum numbers and obey the Bose-Einstein statistics. The Pauli exclusion principle does not hold and two or more pairs may accumulate in one quantum state. 'Many-particle' then means 'many-excitons' created by intense optical excitation and interacting via exchange and polarization interaction (Belleguie and Banyai 1991). The nonlinearity is modelled by the population of states within a system of unbound and bound pair-states (excitons and biexcitons). The theoretical change in absorption is shown in Figs. 3.36 and 3.37 calculated according to this model for different levels of excitation. The nonlinear absorption is caused by a gradual population of the boson states. In the theoretical spectra this fact is reflected by the saturation and the shift of the lowest resonances to higher energies (Fig. 3.36). At higher excitation density a larger number of bosons is generated and transitions occur involving two or more bosons connected with the change in the occupation number larger than one. The boson statistics have to be considered, and a temperature characterizing the boson distribution function is introduced as well as a line broadening taking into account exciton–exciton interaction. As a result, the structures merge into a broader lineshape and a stronger blue shift is obtained (Fig. 3.37). Since the homogeneous linewidth and the confinement induced shift in energy are approximately of the same order of magnitude, the presented results of the calculations made by Belleguie and Banyai (1991) are very sensitive with respect to small changes of line broadening.

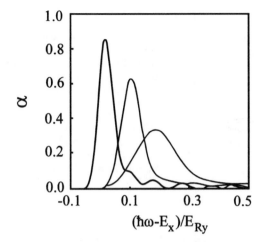

Fig. 3.37. Theoretical absorption spectra considering more than one boson. The curves correspond to $n = 0, 5, 10$ bosons at temperature $T = 0, 100, 200$ K and widths $\Gamma = 0.03, 0.045$, and $0.09 E_{Ry}$, respectively. From Belleguie and Banyai (1991)

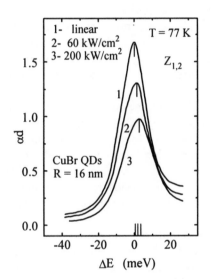

Fig. 3.38. Probe-beam absorption spectra of a two-beam experiment in CuBr quantum dots of $R = 16$ nm ($T = 77$ K) with excitation in the Z_3 exciton state with moderate excitation intensities of 60 (2) and $200 \, kW/cm^2$ (3). Curve (1) is the spectrum without pump. From Jungk and Woggon (1991)

In the experiments reported for semiconductor materials exhibiting center-of-mass confinement, the lowest exciton peak shifts to blue and saturates with increasing optical excitation (Woggon et al. 1988; Masumoto et al. 1989b; Gilliot et al. 1989; Zimin et al. 1990; Jungk and Woggon 1991; Wörner 1992; Gaponenko et al. 1992). Figures 3.38 and 3.39 show the experimental non-linear absorption at CuBr quantum dots with radii of $R = 16$ nm ($\sim 13 a_B$) and CuCl quantum dots with $R = 12$ nm ($\sim 17 a_B$). Figure 3.38 presents a saturation and a small blue shift of the excitonic peak (~ 6 meV) for CuBr quantum dots excited at comparably low intensities of 60 and $200 \, kW/cm^2$ at the spectral position of the Z_3-exciton peak (Jungk and Woggon 1991).

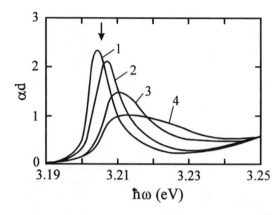

Fig. 3.39. Probe-beam absorption in a two-beam experiment in CuCl quantum dots of $R = 12\,\text{nm}$ ($T = 2$ K) with resonant excitation (pump energy is indicated by arrow) at intensities of 0.07 (2), 1.9 (3), and 5 MW/cm^2 (4). Curve (1) is the spectrum without pump. From Gilliot et al. (1989)

The lowest confined exciton states are hard to resolve. Obviously the very small ideal broadening parameter Γ chosen for the calculations (Fig. 3.36) is not realistic and cannot describe the broader absorption in the quantum dot structures investigated here. However, if taking into account additional mechanisms of line broadening, for example, due to surface scattering, the behavior predicted by theory in Fig. 3.36 can be recognized in Fig. 3.38 as the nonlinear absorption change of CuBr quantum dots at moderate excitation intensities.

Working at larger excitations in the MW/cm^2 region a pronounced blue shift has been observed in CuCl quantum dots [Fig. 3.39 (Gilliot et al. 1989)]. In CuCl the lowest exciton is the Z_3 state followed by the $Z_{1,2}$ state, in contrast to CuBr. The absorption of the Z_3 exciton has been studied under resonant excitation with a spectrally narrow pump laser and maximum intensity of 5 MW/cm^2. Besides the high-energy shift of the absorption maximum, the increasing line broadening can also be seen if the excitation intensity is increased.

When theory and experiment are compared, the calculated spectra resemble the measured ones, but a quantitative discussion about the density necessary to achieve a distinct nonlinearity is difficult at the present stage of experiment and theory. To summarize, different reasons can be held responsible for the observed high-energy shift of the exciton absorption: (i) the population of the first exciton states by a change of the one-pair occupation number, (ii) the contribution of higher pair-transitions to the spectra and to the population or (iii) the influence of an electron–hole plasma and the screening of the exciton.

The latter reason, not yet discussed, means the formation of an electron–hole plasma at high excitation densities. In the plasma state the entire excited volume contains free electron–hole pairs which screen simultaneously the excitonic Coulomb potential. With the increasing degree of excitation of the semiconductor, a gradual decrease of excitonic binding energy occurs and results in a complete bleaching of the excitonic absorption. This effect is known in bulk semiconductors, with the peculiarity that the excitonic blue shift resulting from the change in binding energy is compensated by a red shift caused by the bandgap renormalization. In low-dimensional systems this compensation is weaker and the shift to higher energies becomes observable, for example, for the HH exciton in III–V quantum wells (Peyghambarian et al. 1984).

In I–VII semiconductor materials, the formation of an electron–hole plasma has been detected in CuCl during the first few ps after excitation (Hulin et al. 1985). The dominance of the electron–hole plasma is very unlikely to occur in the range of intensities available without damaging the samples. However, when investigating II–VI quantum dots of large radii, the plasma screening certainly is no longer negligible.

4. Dielectric Effects

In the following chapter we examine possible consequences arising from the fact that a system of lower dimensionality consists of two materials not only with different energy gaps E_g, but also with different dielectric functions $\epsilon(\omega)$. Semiconductor inclusions in a transparent host material represent an optically heterogeneous medium. The optical properties have to be determined by averaging over these local inhomogeneities, introducing suitable effective media models for the composite material. This problem generally occurs for composite materials with semiconductor particles which are small in size but do not necessarily show quantum confinement. The Maxwell-Garnett effective medium theory will be shown to be a good approximation for the optics of ordinary semiconductor doped glasses.

In a first step, we discuss on a macroscopic scale (without consideration of changes of the internal discrete energy states) the influence of the differences in the dielectric functions on the properties of the optical resonances of the quantum dot. In a second step, we present some theoretical predictions from calculations on a microscopic scale concerning the influence of local fields on the confined electronic states. In this case, the polarization energy $\delta V(\epsilon_1, \epsilon_2)$ has been treated as a correction of the Coulomb potential of the quantum dot Hamiltonian (3.19). In small quantum dots with large surface influence, the charge and the polarization state at the interface become essential and determine the intrinsic electronic properties.

4.1 Optical Properties of Composites

The absorption coefficient α_{comp} of an ensemble of embedded quantum dots has been given in (1.1), making use of the simple relation

$$\alpha_{comp} = p\, \alpha_{semicond}, \tag{4.1}$$

where p and $\alpha_{semicond}$ are the volume fraction and the absorption coefficient of the semiconductor material. $\alpha_{semicond}$ is defined by the imaginary part of the dielectric function $\epsilon(\omega)$ of the semiconductor. But more precisely, the dielectric function has to be found for a medium characterized by ϵ_2 embedded in the surrounding host with ϵ_1. The particles will be assumed to be small compared to the wavelength of light so that even for electromagnetic fields

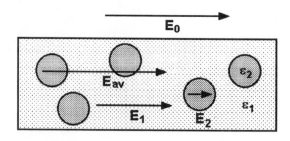

Fig. 4.1. Effective medium with the average dielectric function ϵ_{av} consisting of semiconductor spheres with the dielectric function $\epsilon_2(\omega)$ embedded in a host medium with ϵ_1. E_0, E_1, E_2 and E_{av} are the external electric field, the electric fields in the host medium and inside the sphere and the electric field averaged over a macroscopic volume of the sample

the results of electrostatics provide a good approximation. The electromagnetic field of the incoming light wave polarizes the entities of the composite material. The result is that an averaged dielectric function is observed, which depends on the spectral behavior of ϵ_1 and ϵ_2 itself, on the volume fraction and on the geometrical shape of the constituents of the composites.

The average dielectric function of the composite has to be derived by summing over all the local fields within the microscopic heterogeneous medium. For illustration, in a first step we consider uniformly distributed spheres separated by large distances within the host medium. Homogeneous field conditions have been assumed for the field inside the spheres. While the embedded semiconductor spheres are considered to be absorbing particles characterized by the complex dielectric function $\epsilon_2(\omega)$, the host medium is nonabsorbing, described by the real constant ϵ_1. $\epsilon_2(\omega)$ is supposed to be the same for all semiconductor spheres, i.e. (i) the sizes of the spheres are all identical (constant radius R) or (ii) the energy states are not size-dependent (Fig. 4.1). Therefore the following discussion is going beyond the confinement concept.

The volume fraction p_1 and p_2 occupied by the two constituents relative to the total sample volume are defined as

$$p_{1,2} = \frac{V_{1,2}}{V_1 + V_2} \tag{4.2}$$

with

$$p_1 + p_2 = 1 . \tag{4.3}$$

The field E_1 outside one selected semiconductor sphere is a superposition of the external field E_0 and the dipole fields of all other spheres. The field E_2 inside this sphere is a superposition of E_1 and a depolarization field at the surface of the sphere. For the average electric field and polarization, the relations

$$E_{\mathrm{av}} = p_2 E_2 + (1 - p_2) E_1 \tag{4.4}$$

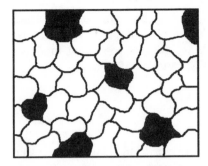

Fig. 4.2. Structure corresponding to the Maxwell-Garnett approximation with $p_2 \ll p_1$ and well-separated particles

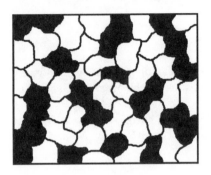

Fig. 4.3. Structure corresponding to the Bruggeman approximation with $p_1 \approx p_2$ and partly aggregated particles

and

$$4\pi P_{\mathrm{av}} = p_2(\epsilon_2 - 1)E_2 + (1 - p_2)(\epsilon_1 - 1)E_1 = (\epsilon_{\mathrm{av}} - 1)E_{\mathrm{av}} \tag{4.5}$$

hold as a crude estimate. The two fields E_1 and E_2 do not act independently from each other. As known from textbooks of electrodynamics (Sommerfeld 1988; Born and Wolf 1959; Bottcher 1973) the field E_2 inside a spherical particle with depolarization factor $1/3$ is

$$E_2 = \frac{3\epsilon_1}{\epsilon_2 + 2\epsilon_1} E_1 . \tag{4.6}$$

In most cases, $\epsilon_1 < \epsilon_2$ (outside of resonances in $\epsilon_2(\omega)$) and the field inside the medium is smaller than outside.

After eliminating the field terms using (4.4–4.6), the discussion yields the following expression of ϵ_{av}

$$\epsilon_{\mathrm{av}}(\omega) = \epsilon_1 \frac{\epsilon_2(\omega)\,(1 + 2p_2) + 2\epsilon_1\,(1 - p_2)}{\epsilon_2(\omega)\,(1 - p_2) + \epsilon_1(2 + p_2)} . \tag{4.7}$$

ϵ_{av} is the effective medium dielectric function. The definition of an effective medium is based on the assumption that the composite material can be replaced by a homogeneous medium described by ϵ_{av} without any change in the optical properties.

An exact treatment has to start from the consideration of the heterogeneous medium as an ensemble of randomly distributed dipoles of densities

N_1 and N_2 and different polarizabilities α_1 and α_2 being in vacuum or in a background dielectricum. The solution of this problem of spatially distributed dipoles and their superimposed local fields results in a somewhat modified Clausius–Mosotti formula (Sommerfeld 1988; Bottcher 1973),

$$\frac{\frac{\epsilon_{av}}{\epsilon_b} - 1}{\frac{\epsilon_{av}}{\epsilon_b} + 2} = \frac{4\pi}{3} \frac{1}{\epsilon_b} \left(N_1 \alpha_1 + N_2 \alpha_2 \right), \tag{4.8}$$

where ϵ_{av} is the dielectric constant of the composite in relation to ϵ_b, the background dielectric constant. If both phases have known dielectric functions, (4.8) can be rewritten as

$$\frac{\epsilon_{av} - \epsilon_b}{\epsilon_{av} + 2\epsilon_b} = p_1 \frac{\epsilon_1 - \epsilon_b}{\epsilon_1 + 2\epsilon_b} + p_2 \frac{\epsilon_2 - \epsilon_b}{\epsilon_2 + 2\epsilon_b}. \tag{4.9}$$

The problem is to find a method for the averaging process which finally leads to different effective medium theories, such as the Maxwell-Garnett or Bruggeman theories.

Within the framework of the Maxwell-Garnett theory (Maxwell-Garnett 1906), phase 2 is assumed to be the diluted phase with $N_2 \ll N_1$ and phase 1 is the host medium. Then $\epsilon_b = \epsilon_1$ and one obtains

$$\frac{\epsilon_{av} - \epsilon_1}{\epsilon_{av} + 2\epsilon_1} = p_2 \frac{\epsilon_2 - \epsilon_1}{\epsilon_2 + 2\epsilon_1}. \tag{4.10}$$

In the case that phase 1 and 2 have comparable volume fractions, the self-consistent choice of $\epsilon_b = \epsilon_{av}$ is commonly made and (4.9) reduces to

$$0 = p_1 \frac{\epsilon_1 - \epsilon_{av}}{\epsilon_1 + 2\epsilon_{av}} + p_2 \frac{\epsilon_2 - \epsilon_{av}}{\epsilon_2 + 2\epsilon_{av}}. \tag{4.11}$$

This is the Bruggeman expression of the effective medium theory (Bruggeman 1935). As illustrated in Figs. 4.2 and 4.3, the microstructural spatial arrangement of the constituents in the composite material is of importance for deriving the dielectric function and the corresponding optical properties. The Maxwell-Garnett theory starts from the assumption of noninteracting spheres of material 2 completely surrounded by material 1. In the framework of the Bruggeman theory, the volume fractions of both materials are of the same order and p_1 and p_2 have the character of probabilities for the constituents to be a small fraction of material 1 or of material 2. Each part of the medium of material 2 or 1 is surrounded by parts of both types of the constituents.

Based on the low filling factor of the semiconductor materials of below 1% (see Sect. 2.4), the semiconductor doped glasses can usually be well described by the Maxwell-Garnett theory. In the following we consider real parameters, i.e. spheres with the volume fraction $p_2 = p$ inside the nonabsorbing glass with the dielectric constant of glass ϵ_1 (usually $\epsilon_1 \approx 2.25$). The semiconductor dielectric function is complex,

$$\epsilon_2(\omega) = \epsilon' + i\epsilon'' , \tag{4.12}$$

and determines the spectral behavior of the semiconductor refractive index and the absorption coefficient. When we insert (4.12) in (4.10), ϵ_{av} is given by

$$\epsilon_{av}(\omega) = \epsilon_1 \, \frac{\epsilon_2(\omega)\,(1+2p) \, + \, \epsilon_1\,(2-2p)}{\epsilon_2(\omega)\,(1-p) \, + \, \epsilon_1\,(2+p)} , \tag{4.13}$$

corresponding to (4.7) from our first-principle discussion.

In the literature concerning the linear and nonlinear optical properties of semiconductor doped glasses, different expressions can be found for the average dielectric function ϵ_{av} of the effective medium semiconductor–glass (Jungk 1988b; Jungk and Woggon 1991; Rustagi and Flytzanis 1984; Hache et al. 1986; Olbright et al. 1987; Chemla et al. 1986; Schmitt-Rink et al. 1987). At least under the assumptions made within the Maxwell-Garnett theory it can be shown that both (4.14) used by Jungk (1988b),

$$\epsilon_{av}(\omega) = \epsilon_1 + \epsilon_1 \frac{3p(\epsilon_2(\omega) - \epsilon_1)}{\epsilon_1\,(2+p) + \epsilon_2(\omega)(1-p)} ; , \tag{4.14}$$

or (4.15) from Jungk and Woggon (1991),

$$\epsilon_{av}(\omega) = \epsilon_1 \left[1 + p \, \frac{\epsilon_2(\omega) - \epsilon_1}{\epsilon_1 + 1/3(1-p)(\epsilon_2 - \epsilon_1)} \right] , \tag{4.15}$$

or (4.16) from Rustagi and Flytzanis (1984),

$$\epsilon_{av}(\omega) = \frac{1+2p}{1-p} \epsilon_1 \; - \; \frac{9p\epsilon_1^2}{(1-p)^2 \left[\epsilon_2(\omega) + \epsilon_1 \dfrac{(2+p)}{(1-p)} \right]} , \tag{4.16}$$

are identical with (4.13), used for example by Olbright et al. (1987).

We are interested to know under which conditions the approximation for the absorption coefficient (4.1) holds, i.e., the simple approximation that the linear absorption spectrum of the composite material is p times the absorption spectrum of the semiconductor spheres. For this purpose we derive from (4.16) the imaginary and real parts of the complex dielectric function $\epsilon_{av}(\omega)$,

$$\mathrm{Im}\,\epsilon_{av} = \frac{9p\,\epsilon_1^2}{(1-p)^2} \, \frac{\epsilon''}{\left(\epsilon' + \epsilon_1 \dfrac{2+p}{1-p} \right)^2 + \epsilon''^2} , \tag{4.17}$$

$$\mathrm{Re}\,\epsilon_{av} = \frac{1+2p}{1-p} \epsilon_1 - \frac{9p\,\epsilon_1^2}{(1-p)^2} \, \frac{\epsilon' + \epsilon_1 \dfrac{2+p}{1-p}}{\left(\epsilon' + \epsilon_1 \dfrac{2+p}{1-p} \right)^2 + \epsilon''^2} . \tag{4.18}$$

ϵ' and ϵ'' are the real and imaginary parts of the semiconductor dielectric function $\epsilon_2(\omega)$. With the assumption $\epsilon' > \epsilon''$ (4.12), only weak resonances are considered in the following. It can then easily be seen that for low filling factors p the spectral behavior of the real part of the dielectric function is determined by the constant value of the refractive coefficient of the nonabsorbing glass ϵ_1. For the absorption coefficient we get

$$\alpha_{\text{composite}} = \frac{\omega}{c \, n_{\text{av}}} \frac{9p \, \epsilon_1^2}{(1-p)^2} \frac{1}{\left(\epsilon' + \epsilon_1 \dfrac{2+p}{1-p}\right)^2 + \epsilon''^2} \epsilon''(\omega) \qquad (4.19)$$

using the relation

$$\alpha = \frac{\omega}{c} \frac{\text{Im}\epsilon}{n} . \qquad (4.20)$$

The approximation $\alpha_{\text{comp}} \approx p \, \alpha_{\text{semicond}}$, (4.1), only holds if

$$
\begin{aligned}
p &\ll 1 \\
n_{\text{av}} &\approx \sqrt{\epsilon'_{\text{av}}} \\
n_{\text{semicond}} &\approx \sqrt{\epsilon'} \\
\epsilon' &\approx \epsilon_1 .
\end{aligned}
\qquad (4.21)
$$

In the usual semiconductor doped glasses, the first three assumptions can be easily fulfilled; the last one is a strong approximation. However, at low filling factors p of the semiconductor material, outside resonances or near weak resonances in the semiconductor dielectric function ϵ_2, (4.1) is a valid approximation for the qualitative discussion of the absorption spectrum of semiconductor-doped glasses.

Similar considerations have been made for the *nonlinear* optical functions of composite materials by Rustagi and Flytzanis (1984), for example, in the case of high laser excitation. The changes in the absorption coefficient and refractive index of the composite arise both from the nonlinear behavior in ϵ' and ϵ'', i.e. from the nonlinear changes $\Delta \epsilon'$ and $\Delta \epsilon''$ of the semiconductor dielectric function. The nonlinear change in the refraction index of the semiconductor determines not only the nonlinear optical behaviour of the average refraction index, but also that of the average absorption coefficient and vice versa.

Whereas the approximation

$$\Delta \epsilon''_{\text{av}} \approx p \, \Delta \epsilon'' \qquad (4.22)$$

is justified for the absorption coefficient (Rustagi and Flytzanis 1984), the nonlinear refractive index of the composite material can be strongly influenced by the nonlinear change in the absorption coefficient of the semiconductor. These nonlinear optical properties of composite materials have been proposed for use in optical bistability based on the intensity-dependent feedback of the nonlinear optical changes (saturating resonances) to the local field factor (Chemla and Miller 1986; Schmitt-Rink et al. 1987; Jungk 1988a; Leung 1986; Li et al. 1989).

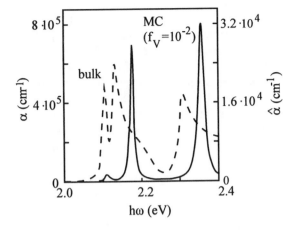

Fig. 4.4. Linear absorption spectrum of a three-resonance model substance (A, B, and C valence-band states) as bulk material (dashed line) and embedded in glass (solid line) calculated with parameters corresponding to bulk CdSSe and $p = 0.01$. From Jungk et al. (1991)

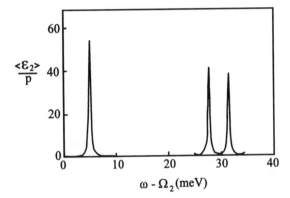

Fig. 4.5. Linear absorption spectrum of a two-resonance model substance (A and B valence-band) after embedding in glass and considering additionally anisotropy, calculated with parameters corresponding to 2 nm CdSe nanospheres. From Ramaniah et al. (1989)

So far, the analysis of the additional boundary condition imposed on the semiconductor sphere when embedded in a dielectric has been performed without consideration of a lowering in dimensionality. It is conceivable that in the case of weak confinement, serious problems exist in the assignment of small shifts in energy of the optical resonances, or variations in oscillator strengths, either to the quantum confinement or to the averaging process because of the differences in the dielectric functions. Indeed, the correction due to the dielectric confinement can reach the same order of magnitude as the confinement-induced energy shift in the special case of two strong, near-neighboring resonances, as will be demonstrated in the following examples.

In Figs. 4.4 and 4.5 theoretical calculations are shown where the absorption coefficients α_{comp} and $\alpha_{semicond}$ do not peak at the same energy (Jungk and Woggon 1991; Ramaniah et al. 1989; Nair and Rustagi 1989) if the semiconductor sphere is embedded into the glass host. The parameters chosen for the calculations correspond to realistic values appropriate to large, bulk-like spherical nanocrystals of CdS_xSe_{1-x} ($x \approx 0.4$) (Fig. 4.4) and CdSe-quantum dots with $R = 2$ nm (Fig. 4.5) as well as volume fractions of $p \leq 0.01$.

Figure 4.4 shows the calculated absorption spectrum of wurtzite-type bulk material with the three valence-band energies A (~ 2.1 eV), B (~ 2.125 eV), and C (~ 2.3 eV) and the changes after incorporation of this semiconductor into glass. A three-oscillator model has been assumed enlarged by a background fundamental gap region and the Urbach tail. Strong changes in the energy positions and relative oscillator strengths occur if the spectra are calculated under consideration of the surrounding glass matrix. Owing to the presence of the neighboring transition in the optical spectrum an enhancement in both the strength and the shift of the upper resonances and a suppression of the lower resonances can be seen after the embedding. Although this effect' imitates a confinement-induced blue shift, it is only the result of dielectric properties.

A two-resonance model has been used by Ramaniah et al. (1989), and Nair and Rustagi (1989) to study the effect of anisotropy in CdS and CdSe quantum dots of wurtzite-type structure. As seen in Fig. 4.5 for CdSe quantum dots (bulk A–B splitting 25 meV), not only a shift in the lowest resonance of about 5 meV has been obtained, but also a splitting of the B-resonance of 3.7 meV due to the anisotropic semiconductor dielectric function. Compared with the experimentally observed confinement-induced energy shift of about 160 meV for 2 nm CdSe quantum dots, the dielectric effects are negligible in this case. Thus, for CuBr quantum dots, the beginning of quantum confinement, indicated usually by energy shifts of a few meV only, could also be simulated by shape variations or changes of the damping parameters only (Jungk and Woggon 1991). For CuBr quantum dots with $R \geq 10a_B$ the effective medium analysis of the shifts in the optical spectra does not show any sure hint of quantum confinement. In this case one has to be careful to extract size effects from experimental data without exact knowledge of the volume fraction and shape of the nanocrystals.

The experimental observation of these theoretical predictions has not yet been possible. It requires identical, uniformly shaped nanocrystals with sharp resonances. However, one has to keep in mind the existence of these or related effects if interpreting the spectra under conditions mentioned above, i.e., weak confinement, strong optical transitions, higher filling factors, and near neighboring resonances.

4.2 Surface Polarization and Charge Separation

The inherent difference in the dielectric functions at the boundary semiconductor–matrix is the reason for the formation of *induced* or *image* charge distributions. By, for example, optical excitation within this effective medium, additional *external* electron–hole pairs can be created and, owing to the presence of point charges, an additional inhomogeneous field is superimposed. Therefore, the electron–hole pair energy states and wave functions are expected to be modified in the presence of interfaces between the two media of different dielectric constants. The Coulomb problem of the electron–hole pair in the sphere has to be solved under consideration of the induced charge distributions, partly screening the bare Coulomb attraction and leading to a corrected effective Coulomb interaction potential. The corresponding Hamiltonian (Brus 1984; Brus 1986) describing the dielectric contribution to the confinement is given by

$$
\hat{H} = -\frac{\hbar^2 \nabla_e^2}{2m_e} - \frac{\hbar^2 \nabla_h^2}{2m_h} - \frac{e^2}{\epsilon_2 |r_e - r_h|}
$$
$$
+ \frac{e^2}{2R} \sum_{n=0}^{\infty} \alpha_n \left[\left(\frac{r_e}{R}\right)^{2n} + \left(\frac{r_h}{R}\right)^{2n} \right]
$$
$$
- \frac{e^2}{R} \sum_{n=0}^{\infty} \alpha_n \left[\frac{r_e r_h}{R^2} \right]^n P_n(\cos \Theta_{eh}) + V_e(r_e) + V_h(r_h) . \tag{4.23}
$$

Here r_e, r_h, m_e, m_h are the electron and hole coordinates and masses, respectively, P_n is the Legendre polynomial of nth order, and $\Theta_{e,h}$ is the angle between electron and hole coordinates. α_n is defined by

$$
\alpha_n = \frac{(n+1)(\epsilon - 1)}{\epsilon_2 (n\epsilon + n + 1)} \tag{4.24}
$$

with $\epsilon = \epsilon_2/\epsilon_1$. The two new terms define the surface polarization energy resulting from (i) the interaction of the electron and the hole with their own image charge (fourth term) and (ii) the mutual interaction between electron and holes via the image charges (fifth term). The effect of the penetration of an electric field into the surrounding matrix due to the smaller dielectric constant outside the sphere results in a correction of the Coulomb interaction between electron and hole.

An analysis of the electrostatic problem of additional point charges inside a sphere of ϵ_2 embedded in ϵ_1 has been carried out by Brus (1984, 1986). To determine the field at an arbitrary point inside the sphere, the Poisson equation has been solved assuming infinitely high potentials with the boundary condition of vanishing wave functions outside the sphere.

The effect of dielectric confinement on the pair-state energies has been analyzed by use of variational methods by Takagahara (1993a). Assuming infinite potential barriers and the strong confinement case, an enhancement of the pair-state binding energy has been found with increasing ratio ϵ_2/ϵ_1 as shown in Fig. 4.6. For the case of CdSe quantum dots embedded in glass, an estimate of about ~ 3 to ~ 4 has been given for the ratio of ϵ_2/ϵ_1, which stands for a significant energy correction due to the dielectric confinement.

In the framework of the matrix diagonalization method, the one- and two-pair states have been treated including the surface polarization $\delta V(r_e, r_h, \epsilon_1, \epsilon_2)$

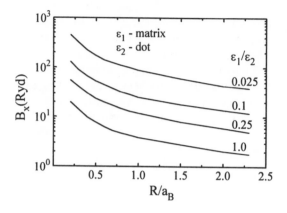

Fig. 4.6. The binding energy of the lowest exciton state plotted as a function of the dot radius with ratio of the dielectric constants as the parameter, m_e/m_h = 0.2. From Takagahara (1993)

by Hu et al. (1990b), again under consideration of infinitely high barriers. As long as the barrier potential is supposed to be infinite, the discontinuity of the electric field at the boundary between the two media is not significant with respect to microscopic energy changes, in contrast to the more realistic case of finite barriers. In the case of $\epsilon_2 > \epsilon_1$ (4.6), the potential just outside the sphere becomes attractive for a charged particle within an infinitesimal small distance $R^+ > R$. At very low barrier heights this potential minimum outside the sphere would capture all charged particles. It is expected that a scenario develops where the capture probability is determined by the interplay between the barrier height and the potential minimum resulting from the differences in the dielectric constants inside and outside the dot.

The problem of the electron–hole pair ground state in a quantum dot involving both finite barriers as well as interface dielectric effects has been treated by Banyai et al. (1992). The singularity at the boundary has been overcome by introducing a phenomenological cut-off length at the interface, which excludes a layer in the order of magnitude of the lattice constant and therefore regularizes the potential. The two-particle Hamiltonian is similar to (3.25) and involves the corrected Coulomb potential due to the dielectric effects together with the regularization at the interface, however now with finite V_e and V_h.

The interesting result obtained after calculating the energy states and wave functions with the barrier height as the parameter (at fixed size $R = a_B$ and $\epsilon_2/\epsilon_1 = 10$) is shown in Fig. 4.7. In particular, the maximum of the hole wave function gradually shifts from the center of the sphere (at high parameter values for the barrier potential) towards the interface when the barrier potential V_0 decreases. Dependencing on the barrier height and the mismatch in the dielectric constants, a so-called volume state (hole in the center) develops to a surface state (hole localized at the interface). (The ideal volume state corresponds to the radial distribution in the case of infinite barriers $V_0 \longrightarrow \infty$). Simultaneous to the localization at the interface, a substantial

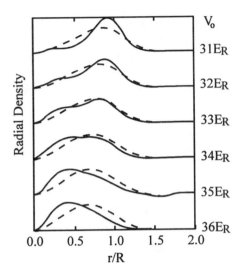

Fig. 4.7. Radial distribution of electrons (dashed) and holes (solid) obtained within the matrix diagonalization method for different potential barrier heights V_0 (in Rydberg units) and $R = a_B$, $\epsilon_2/\epsilon_1 = 10$ and $m_e/m_h = 0.1$. From Banyai et al. (1992)

modification in the energy states occurs, leading to a shift to lower energies. The red-shift of the absorption compensates the confinement-induced blue-shift and the real energy shift is always the sum of both contributions.

Although in quantum dots there is neither the situation of abrupt interfaces within a monolayer distance, nor is the height of the potential barrier exactly known, the theoretical results presented by Banyai et al. (1992) are very helpful to understand a number of experimental results observed in small quantum dots.

The small shift of the maximum in the linear absorption observed for quantum dots with $R \leq a_B$ after changing their surface configuration (for example due to hydrogenation or due to growth processes with different ligand groups) has already been attributed intuitively to charge states changes at the interface (see Sect. 2.4). The sign of the shift is difficult to discuss because of the compensation of the confinement-induced blue-shift and the dielectric red-shift mentioned above.

The calculations made by Banyai et al. (1992) support these assumptions as well as the idea of an enhancement of surface trapping in small quantum dots, including the possibility of carrier escape out of the sphere into the matrix (see Chap. 6) or the hypothesis of increasing coupling between electron–hole pairs and LO-phonons (see Chap. 5).

Thus, in small semiconductor quantum dots, surface-related charge separation and polarization effects could be the reason for a shift of the energy states and an increase in the homogeneous line broadening. Since the published data of differential absorption experiments differ substantially in part, it can be supposed that there is a continuous range for the development of states more similar to volume- or to surface-states depending on the interface properties. Obviously, the radial probability density of the electron- and hole wave-function is very sensitive with respect to growth parameters in the case of quantum dots of small sizes.

In conclusion, we have seen that especially in the strong confinement limit, it is necessary in the first place to clarify which kind of states are determining the electronic properties before opening the discussion of experimental results. It is useful to investigate quantum dots exhibiting stable interface configurations. The aim of growth optimization should be directed at a minimization of the differences, in part huge, between the ideal spheres considered in the models and the nanocrystals which really exist.

5. Mechanisms of Dephasing

In the preceding chapters the discussion of the confinement-induced changes has concentrated so far on the *weights* and the energy *positions* of the optical transitions. The complete description of the observed absorption spectra needs the additional information about the mechanisms determining the homogeneous *line-width*. The scope of this chapter is the investigation of the mechanisms of dephasing in semiconductor quantum dots. The term 'phase relaxation' (or 'dephasing') comprises all processes bringing back coherently driven dipoles excited by the light field to randomly distributed polarizations. The simplest model to introduce the phase relaxation are the optical Bloch equations formulated for a two-level system. Although the correct understanding of an optically excited bulk semiconductor requires a many-body formalism (Haug and Koch 1993), the simple two-level system should at least provide qualitatively a good description of the relaxation dynamics in quantum dots with their discrete level structure.

The optical Bloch equations and the density matrix formalism combine Maxwell's equation of the electromagnetic waves with the time-dependent evolution of polarization and population of the medium. This treatment goes beyond the classical anharmonic oscillator model and starts with a semiclassical ansatz (see Bloembergen 1977; Shen 1984; Yariv 1989; Meystre and Sargent 1990; Schubert and Wilhelmi 1978). The dynamics of such a system of non-interacting two-level atoms are completely determined by the population lifetime T_1 and the phase relaxation time T_2. The time T_1 is also conventionally termed the longitudinal and T_2 the transversal relaxation time, respectively. The relation

$$\frac{1}{T_2} = \frac{1}{2T_1} \tag{5.1}$$

holds in the case of pure dephasing due to the finite lifetime of the excited state. In general, several scattering mechanisms are involved in the phase relaxation dynamics giving a sum over the different T_i^*,

$$\frac{1}{T_2} = \frac{1}{2T_1} + \sum_i \frac{1}{T_i^*} . \tag{5.2}$$

T_2 can be correlated to the homogeneous line-width Γ_{hom} of the optical transition by

$$\Gamma_{\text{hom}} = \frac{2\hbar}{T_2} \,. \tag{5.3}$$

In the experiments one tries to find conditions where only one process dominates the pure dephasing. To investigate the loss of phase coherence in semiconductors (and their nanostructures), ultrafast lasers have been used in degenerate four-wave mixing, photon echo or quantum beat experiments. Likewise, experimental results obtained by nonlinear optical methods working in the frequency domain give insight into the dephasing processes, such as spectral hole burning and non-degenerate four-wave mixing by use of nanosecond dye lasers. In analogy to the bulk we take over the distinction of the phase destroying processes in (i) coupling to different types of phonons, (ii) scattering at defects or interfaces, and (iii) scattering within a many-particle system of interacting electrons and holes. Additionally we have to consider phase destroying energy relaxation. This process is strongly influenced by the mixing of the valence-band wavefunctions as well as the density of free final states for the scattering process.

Scattering of electron–hole pairs in systems of reduced dimensionality by phonons, free carriers or defects is of fundamental interest in semiconductor physics, and also of interest in ultrafast semiconductor devices and lasers. The mechanisms of dephasing in zero-dimensional systems are a central point of present research work. Present knowledge is far from full understanding and in the following we present some first results of preliminary character obtained from experiments carried out mostly on II–VI quantum dots.

Because the T_1 and T_2 processes can develop on similar time scales, Chap. 5 and the following Chap. 6, in which we will study the recombination processes, are closely related.

5.1 Coupling of Electron–Hole Pairs with Phonons

5.1.1 Phonons in Quantum Dots and Coupling Mechanisms

The analysis of the interaction of electron–hole pairs with phonons starts with the investigation of the two main contributions arising from (i) the coupling to acoustic phonons by deformation potential or piezoelectric interaction and (ii) the polar Fröhlich coupling to optical phonons. The Fröhlich coupling scheme describes the coupling of the electric field created by the vibrations of the ionic nuclei with the Coulomb field of the optically excited electron–hole pairs. Due to the deformation potential, however, excitons couple to longitudinal acoustic phonons. The compression and expansion changes the ion bond length thus modulating the semiconductor bandgap and the exciton energy. Via piezoelectric interaction, again a polar coupling scheme, the transverse acoustic phonon mode couples to the exciton.

In bulk II–VI semiconductors the free exciton is most strongly coupled to optical phonons by polar Fröhlich interaction. The deformation potential and piezoelectric coupling of electrons and holes to phonons are usually less important in bulk II–VI compounds. In quantum dots, the problem depends very sensitively on the explicit form of the electron–hole wave functions. In a spherical potential with infinite barriers and when the Coulomb interaction is neglected, the wavefunctions of electrons and holes are identical and no differences in their charge distributions exist. Therefore, at first sight, a reduction of the coupling to LO-phonons has been supposed when further lowering the dimensionality (Schmitt-Rink et al. 1987). However, because of the small dimensions, a significant increase in the coupling to low-energy acoustic phonons might occur. Thus, for quantum dots one has to answer the questions about the existing types of phonons and the corresponding coupling mechanisms.

In the following, the polar coupling scheme and surface phonons will be discussed, and new types of acoustic modes and their coupling to pair states will be considered.

Optical Phonons. In the case of polar interaction, the optically excited electronic charge distribution couples to the field, induced by the vibrational motion of the lattice atoms. If the bulk phonon properties are maintained in the quantum dot, the actual strength of the polar coupling depends on the radial distribution of the wavefunctions considered for the confined electrons and holes. In recent theories (see Chap. 3) some reasons have been proposed giving rise to differences in the radial parts of the wavefunctions. At first, the consideration of finite barriers and the inclusion of the Coulomb interaction in the Hamiltonian disturbs the wavefunctions of electron and hole and causes small differences in their radial distribution. Besides, more pronounced differences in the wavefunctions can be achieved considering an additional charge inside the quantum dot. Differences in the radial part of the wavefunction can also be obtained by taking into account the confinement-induced valence-band mixing.

Figure 5.1 presents an example for a radial charge distribution when including the real valence-band structure into the hole Hamiltonian (Efros 1992b). Three charge regions can be seen, two positive ones at the center and near the interface $R = r$ and a negative one between them. By Nomura and Kobayashi (1994), an exciton with a massive hole trapped at an arbitrary position in the quantum dot has been studied theoretically and the energy minima as well as the density of states and the dipole moment have been calculated. The dipole moment depends strongly on the assumed position of the hole in the quantum dot as well as on the size of the dot.

The phenomenon of intrinsic charge distributions given by the explicit electron and hole-wave functions in quantum dots implies an increase in electron–phonon interaction with decreasing sizes. However, such polarity in the pair-wave function can also have extrinsic reasons. Defects near the

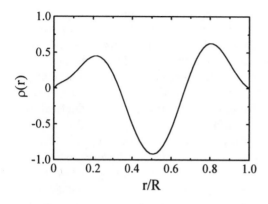

Fig. 5.1. Radial charge distribution in a quantum dot with one electron–hole pair calculated by Efros (1992b) under consideration of the real valence-band structure

semiconductor–glass interface creating internal charge states have been likewise proposed as being responsible for a polarity of the electron–hole-pair wave function. The various attempts to describe the electron–hole pair phonon coupling theoretically, in fact differ only in the structure of the Hamiltonian used. This finally determines the charge distribution in the quantum dot given by the calculated wavefunctions. The various extrinsic and intrinsic reasons for changes in the charge distribution inside the quantum dots can lead to a 'tunable' coupling strength among the existing samples originating from different manufacturing processes and having different semiconductor-matrix interface configurations.

One experimental approach to study the electron–hole pair phonon interaction is the analysis of resonant Raman scattering as well as of the Stokes shift between absorption and luminescence peak. Qualitative discussions can be made in terms of the Huang–Rhys parameter S (Huang and Rhys 1950) comparing the ratio between the zero-phonon line and the corresponding higher phonon satellites (see Fig. 5.2). This approach is based on simple phenomenological models resulting from the Franck–Condon picture with the ground and the excited states being displaced harmonic oscillators. The line shape in absorption and luminescence, and the cross section of Raman scattering can be derived, if calculating the overlap integral of the wavefunctions of the undisturbed and the displaced harmonic oscillators.

In absorption and luminescence according to Huang and Rhys (1950), the optical phonons and their set of satellites give rise to a line shape of the spectra corresponding to

$$\alpha(\omega) = A \exp\left[-S(2\bar{n}+1)\right] \tag{5.4}$$

$$\times \sum_{p=-\infty}^{\infty} \left(\frac{\bar{n}+1}{\bar{n}}\right)^{\frac{p}{2}} I_p(2S\sqrt{\bar{n}(\bar{n}+1)}) \, \delta(E - E_x + S\hbar\omega_{LO} - p\hbar\omega_{LO})$$

with $\bar{n}$ the thermal average of the vibrational quantum number

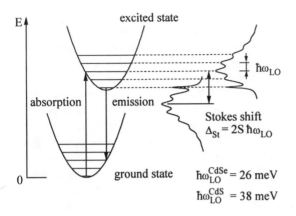

Fig. 5.2. Scheme of the Franck–Condon model (see text)

$$\bar{n}(T) = [\exp(\hbar\omega_{LO}/kT) - 1]^{-1} . \tag{5.5}$$

S is the so-called Huang–Rhys factor, I the Bessel function of second kind, ω_0 the LO-phonon frequency and p the number of the phonons taking part in the optical transitions. Equation (5.4) represents a series of equally spaced delta-functions with the energy separation of the phonon energy. At zero temperature, (5.4) corresponds to the Poisson distribution $S^l e^{-S}/l!$. For $S < 1$, the maximum of the envelope over all phonon replica is situated near the zero-phonon line; in the case of $S > 1$ the phonon satellites are stronger in their intensity compared to the zero-phonon line and the maximum of the envelope shifts to lower/higher frequency for phonon emission/absorption, respectively. The difference between the maximum of absorption and emission, the so-called Stokes shift in the Franck–Condon model , is given by

$$\Delta_{Stokes} = 2 \, S \, \hbar\omega_{LO} \tag{5.6}$$

and the line-broadening due to the phonon coupling is given by

$$\Delta_{FWHM} = \hbar\omega_{LO} \, 2\sqrt{S} . \tag{5.7}$$

In the Franck–Condon model the displacement δ of the minima of the two parabola describing the vibronic potentials and representing the degree of vibrational coupling, is related to the Huang–Rhys Parameter by (see e.g. Huang and Rhys 1950, Banyai and Koch 1993)

$$S = \frac{\delta^2}{2} . \tag{5.8}$$

A first theroretical and experimental investigation of the Huang–Rhys parameter S and, thus, of the electron–phonon coupling strength in quantum dots has been given by Klein et al. (1990) considering uncorrelated electrons and holes. Because of the lack of knowledge of the exact wavefunctions, the authors use a model charge distribution where only the electron is confined and the hole resides at the center of the sphere. For such a charge distribution a size-independence of the electron–phonon coupling has been derived from the calculation. Good agreement between the calculations and experimental results from resonant Raman scattering has been achieved when choosing the Huang–Rhys parameter in the order of $S \sim 0.5$.

The exciton–LO-phonon coupling in three-dimensionally confined spherical semiconductors has also been treated theoretically by Nomura and Kobayashi (1992a). Here various contributions have been taken into account for the precise determination of the wave functions, namely the nonparabolicity of the conduction band, the valence-band mixing and Coulomb interaction. A non-monotonic size-dependence of the exciton–phonon coupling has been obtained, which is closely related to the spatial extension of the exciton wave function itself. For CdSe quantum dots, the coupling of the exciton to LO phonons has a minimum around $R = 7\,\mathrm{nm}$ ($a_{\mathrm{B}}^{\mathrm{bulk}} = 5.6\,\mathrm{nm}$). It becomes stronger in the range $R < 7\,\mathrm{nm}$, mainly due to the confinement-induced mixing of hole wave functions. In the range $R > 7\,\mathrm{nm}$ the confinement is more influenced by the Coulomb interaction, and again the Huang–Rhys parameter of the exciton increases with increasing radius. S has been calculated with and without an extra charge at the center of the quantum dot. In the first case an increase in S up to values around 1 can be reached, which gives rise to the conclusion that the S-values observed in the experiments (see below) might originate from the presence of charged point defects.

As already outlined in Chap. 4, a further ansatz has been proposed by Banyai et al. (1992). It is suggested that the discontinuity in the dielectric constants at the interface between the two media modifies the effective Coulomb interaction and results in a stronger coupling between electron–hole pairs and LO phonons because of the localization and self-trapping of the hole near the surface, whereas the electron remains delocalized within the quantum dot.

A model introducing a more general application of the Franck–Condon principle has been proposed by Sercel (1995). The author couples the electron–hole pair of the dot with a deep trap level located in the surrounding matrix. Fast, multiphonon-assisted energy relaxation is the result of this mechanism. We come back to this model in Chap. 6.

From the point of theory it seems that theoretical predictions give values of S not exceeding 0.1–0.5, also if taking into account spin–orbit interaction, finite potential wells, nonparabolicity and Coulomb interaction. S around (or larger) than 1 could only be attained by artificially introducing a strong charge separation, for example, by an additional charge (Nomura and Kobayashi 1992a), by surface polarization (Banyai et al. 1992, 1993), or by other effects breaking the spherical symmetry such as strain and lattice deformation. The predicted size-dependence of the coupling strength of electron–hole-pairs to LO phonons differs considerably in the published results; both constancy, and increase and decrease with increasing confinement can be found.

Until now, for discussion of the coupling between electron–hole pairs and LO phonons, we have only regarded the well-known bulk optical phonon modes. A further source of polar optical phonon modes lies in the fact that a spherical entity of dielectric function $\epsilon_1(\omega)$ is embedded inside a medium with $\epsilon_2(\omega)$. This peculiarity gives rise to polar modes which do not exist in an infinite homogeneous medium. Besides the bulk transversal optical phonon ω_{TO}, the bulk longitudinal optical phonon with $\omega_{LO} = \omega_{TO}\sqrt{\epsilon_0/\epsilon_\infty}$, a third type of optical phonons exists in such a sphere, the SO-surface mode with ω_{SO} between the LO- and TO-mode frequencies (ϵ_0 and ϵ_∞ are the bulk static and high-frequency dielectric constants). The expression for the surface mode frequency can be derived from the classical continuum theory of diatomic crystals (without retardation) by paying additional attention to the shape of the dielectric structure (see e.g. Ruppin 1975; Knipp and Reinecke 1992). When we consider the standard equations of electrodynamics

$$D = \epsilon E \qquad\qquad E = -\nabla\phi \qquad\qquad \nabla D = 0 \qquad (5.9)$$

for the dielectric displacement D, the electromagnetic field E and the electrostatic scalar potential ϕ, the basic relation

$$\epsilon_i(\omega)\Delta\phi = 0 \qquad (5.10)$$

can be obtained, which at first holds for each of the materials i. Each material is characterized by its own dielectric function

$$\epsilon_i(\omega) = \epsilon_{\infty,i}\frac{\omega_{LO,i}^2 - \omega^2}{\omega_{TO,i}^2 - \omega^2}, \quad i = 1, 2 . \qquad (5.11)$$

From (5.11) we see that the condition $\epsilon_i(\omega) = 0$ describes a bulk LO-mode and $\epsilon_i(\omega) = \infty$ gives a bulk TO-mode in the material i. To fulfill the condition (5.10) for the interface mode, defined by the properties of *both* materials resulting in $\epsilon_i(\omega) \neq 0$ in general, we have to solve the Laplace equation $\Delta\phi = 0$ with the standard electrostatic boundary conditions imposed at the interface between the media, which are the continuity of the tangential components of E and of the normal component of D. To express the electrostatic potential ϕ, the eigenfunctions have to adopt the spherical symmetry of the sample. The angular functions involved are spherical harmonics and the radial functions are spherical Besselfunctions (Ruppin 1975, Klein et al. 1990, Knipp and Reinecke 1992). When the boundary conditions are applied, the eigenfrequencies are then given by the condition

$$\frac{\epsilon_1(\omega_{lm})}{\epsilon_2(\omega_{lm})} = -1 - \frac{1}{l} . \qquad (5.12)$$

Since for the glass matrix a constant dielectric function can be assumed with

$$\epsilon_2(\omega) = \epsilon_M , \qquad (5.13)$$

we obtain with (5.11) for the surface mode frequencies

$$\omega_{SO} = \left[\frac{\epsilon_{\infty,1}\, l\, \omega_{LO,1}^2 + \epsilon_M\, (l+1)\, \omega_{TO,1}^2}{l\, \epsilon_{\infty,1} + \epsilon_M\, (l+1)} \right]^{1/2} \tag{5.14}$$

with $\epsilon_{\infty,1}$ the high-frequency dielectric constant and $\omega_{LO,1}$, $\omega_{TO,1}$, the vibration frequencies of the LO-, and TO-phonons of the semiconductor material. An example has been given for CdSe quantum dots embedded in glass by Klein et al. (1990) calculating the SO-mode frequencies for $l = 1, 2, \ldots \infty$ to $194\,\mathrm{cm}^{-1}$, $197\,\mathrm{cm}^{-1}$, ... $200\,\mathrm{cm}^{-1}$ with $\epsilon_M = 2.25$, $\epsilon_\infty = 6.1$, $\omega_{LO} = 210\ \mathrm{cm}^{-1}$ and $\omega_{TO} = 170\,\mathrm{cm}^{-1}$. The new aspect of surface (or interface modes) appearing in spherical microcrystals has also been discussed by Hayashi and Kanamori (1982) for GaP spheres, by Pan et al. (1990) and Chamberlain et al. (1995) for CdS, by Marini et al. (1994) for CdSe and CuCl, and by de la Cruz et al. (1995) and Roca et al. (1994) for GaAs microcrystals (see also Sect. 5.1.2).

Acoustic Phonons. Without doubt, the acoustic phonons gain a great deal of attention in quantum dots because of the interesting fact of a supposed decoupling of the electron–hole pair from the phonon bath when the confinement-induced level separation exceeds the LO-phonon energy. Thus energy relaxation processes should become more difficult in quantum dots due to the lack of energy-matched levels. The role of acoustic phonons in the process of energy relaxation has been investigated by Bockelmann and Bastard (1990), Nojima (1992), Bockelmann (1993), and Benisty (1995), and we will come back to this topic in Sect. 5.2.

In quantum dots new types of spherically confined acoustic phonons have been proposed, for example, in theoretical works of Nomura and Kobayashi (1992b) and Takagahara (1993b). For the energies of these acoustic modes a typical $1/R$-size dependence has been found. To illustrate these acoustic vibrations, the model of a spherical elastic body can be used (see e.g. Tamura et al. 1982; Tanaka et al. 1993). The eigenfrequencies can then be obtained by solving the differential equation

$$\partial^2 u/\partial t^2 = \frac{\lambda + \mu}{\rho}\,\mathrm{grad\ div}\ u + \frac{\mu}{\rho}\nabla^2 u \tag{5.15}$$

with u the lattice displacement vector, ρ the mass density and the two parameters μ and λ are Lamés constants. Equation (5.15) is based on the macroscopic elastic continuum model applied to a sphere (Lamb 1982; Stephani and Kluge 1995). This model is valid if the wavelengths of the acoustic phonons are sufficiently larger than the lattice constants of the semiconductor. According to this theory, the constants μ and λ connect the diagonal and off-diagonal elements of the deformation tensor ϵ_{ik} with that of the strain tensor σ_{ik} as

$$\sigma_{ik} = 2\,\mu\,\epsilon_{ik} + \lambda\,\delta_{ik}\,\epsilon_{ll}\ . \tag{5.16}$$

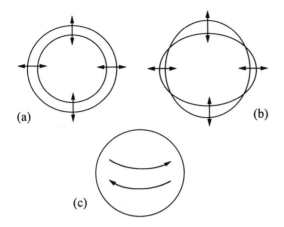

Fig. 5.3. Oscillation modes of a spherical elastic body. (a) 'breathing' mode ω_S with $l = 0$, $n = 1$, (b) an example for a spheroidal mode ω_S with $l \geq 0$, $n = 0$, (c) an example for a torsional mode ω_T with $l = 1$, $n = 0$

With the help of μ and λ, such macroscopic values can be defined as the constant of elasticity, the shear constant, the Poisson constant of cross contraction, or the compression constant. Likewise the longitudinal and transversal velocities of sound in solids can be expressed with these constants as

$$c_\ell = \sqrt{\frac{2\mu + \lambda}{\rho}} \qquad\qquad c_t = \sqrt{\frac{\mu}{\rho}} . \qquad\qquad (5.17)$$

Solving (5.15) under stress-free boundary conditions at the sphere surface, two types of vibrational modes can be obtained, the spheroidal modes ω_S and the torsional modes ω_T. The frequencies of the spheroidal modes depend on the material via the ratio of the longitudinal and transversal sound velocities c_ℓ/c_t. The torsional mode is a vibration without changes in the volume of the sphere, i.e., without dilation and compression.

The eigenvalues are written as follows

$$\xi_S^l = \frac{\omega_S^l R}{c_\ell} \qquad\qquad \eta_S^l = \frac{\omega_S^l R}{c_t} \qquad\qquad (5.18)$$

for the spheroidal modes, and for the torsional modes

$$\eta_T^l = \frac{\omega_T^l R}{c_t} \qquad\qquad\qquad\qquad (5.19)$$

with R the radius of the sphere and l the angular momentum quantum number. Besides the quantum number l, the eigenmodes are characterized by n, giving the order of the zeroth of the radial part of the wave function. The torsional mode is only defined for $l \geq 1$ because the mode with $l = 0$ has no displacement. Oscillation modes with $n = 0$ are characterized by large amplitudes near the surface of the sphere. Eigenfrequencies with $n \geq 1$ correspond to inner modes. The eigenvalues ξ_S^l, η_S^l and η_T^l have to be determined in relation to the semiconductor material. Typical acoustic modes of an elastic sphere are illustrated in Fig. 5.3. Since ξ_S^l, η_S^l and η_T^l are simple numbers, the

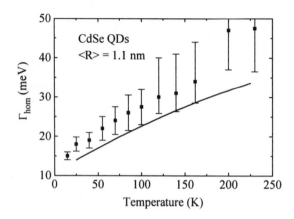

Fig. 5.4. Homogeneous line broadening Γ_{hom} in relation to temperature calculated by Takagahara (1993b). Γ_{hom} is plotted for the lowest pair state of a CdSe quantum dot with 1.1 nm radius together with experimental data obtained by Schoenlein et al. (1993)

eigenfrequencies ω_{S}^l and ω_{T}^l, i.e., the energies of these spherically confined acoustic modes show a $1/R$-dependence on size (5.18, 5.19). The absolute values of the energies amount to a few meV.

The interaction between electron–hole pairs and acoustic phonons arises mainly from the deformation potential coupling. The size dependence of this coupling has been examined for the energetically lowest pair state in quantum dots, for example, by Nomura and Kobayashi (1992b) and Takagahara (1993b). The strength of the deformation potential coupling to the above described confined acoustic phonons has been derived as $\sim 1/R^2$ and thus increases with decreasing size of the quantum dots. Although the II–VI bulk material has large piezoelectric coefficients, the piezoelectric coupling has been found to be of minor importance in the small size regime.

To compare the theoretical results with experiments, Takagahara (1993b) has calculated the temperature dependence of the homogeneous line broadening for CdSe quantum dots. Figure 5.4 shows the result for the lowest pair state when only considering deformation potential coupling. It has been compared with experimental data obtained by Schoenlein et al. (1993) (see also Sect. 5.1.4). Satisfactory agreement can be seen.

5.1.2 Raman Scattering

Raman scattering offers the possibility to determine the relevant phonon energies, to investigate the coupling strength and to examine the growth process. Examples for the usefulness of Raman data for characterizing the grown structure of the crystallites are the experiments by Miyoshi et al. (1995), which derive the composition x of CdS_xSe_{1-x}-mixed crystals from Raman measurements, and the proof that nanocrystals embedded in glass can suffer compressive strain by Scamarcio et al. (1992). The reason for involving strain for the understanding of the Raman data was the unexpected blue shift of the peak position of the LO-phonon energy. Low-frequency Raman scattering

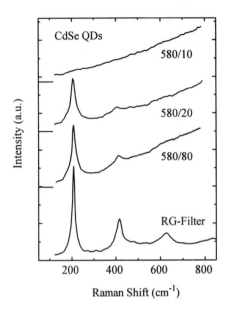

Fig. 5.5. Raman spectra of CdSe quantum dots, grown at 580 °C for 10, 20 and 80 h showing the development of the LO-phonon replica. The Raman spectrum of a RG630-filter glass is added for comparison. From Trinn et al. (1991)

has been used both to determine the radius of the crystallites (Champagnon et al. 1993) and to test the onset of the nanocrystals growth when gradually increasing the temperature (Roy and Sood 1996).

Figure 5.5 shows the results of resonance Raman scattering at CdSe-doped glasses grown after the heat treatment described in Sect. 2.2. The measurements demonstrate the development of the crystalline lattice during the heat treatment. The Raman spectra are plotted for the growth conditions 580 °C/10, 20 and 80 h, indicating the gradual development of the phonon replica during the formation of the crystallites from molecules or large clusters (Trinn et al. 1991). The size of crystallites in the sample grown at 580°C/80 h is about $R = 1.5$ nm; the other glasses contain correspondingly smaller nanocrystals not measurable by usual small-angle X-ray scattering. For comparison the Raman spectrum of the commercial RG630 filter with considerably larger crystallites is added in Fig. 5.5.[1] The phonon side bands, as replicas of the zero-phonon energy, appear in the Raman spectra simultaneously with the development of the crystal lattice. After finishing the construction of the lattice structure in larger-sized quantum dots (sample 580/80 and RG630 filter glass), it is reasonable to discuss the ratio between the zero-phonon intensity and the phonon replica in the framework of the Huang–Rhys theory. For the analysis of the Raman cross section, similar equations can be used like those discussed for the photoluminescence (5.4)–(5.8). More details can be found,

---

[1] Nominally the RG630 filter contains CdS_xSe_{1-x}-mixed crystals, but x is very small and no CdS-phonon mode has been observed in the Raman spectra

Table 5.1. Size dependence of the coupling parameter S for CdS quantum dots in different matrices: g: glass, phe: thiophenol-stabilized, gly: thioglycolate-stabilized, pho: polyphosphate-stabilized. S has been calculated assuming a constant line broadening of $50\,\mathrm{cm}^{-1}$. Data taken from Shiang et al. (1993)

Radius (nm)	Matrix	0ph/1ph ratio	S
0.9	g	5	<0.1
1.5	pho	2.17	0.4
1.6	phe	2.32	0.3
1.6	gly	2.44	0.3
1.65	gly	3.12	0.2
3.4	gly	1.66	0.5
3.5	gly	1.43	0.7
3.5	pho	1.33	0.8
3.6	phe	1.29	0.9
5	gly	1.4	0.7
6.4	gly	1.08	>1.4

for example, in Klein et al. (1990), Merlin et al. (1978), Cardona (1982), and Champion et al. (1979). It should be mentioned that the intensity ratio between the zero-phonon line and the different satellites depends not only on S or δ, but the line-broadening Γ is also involved in the calculations. Usually, this parameter has been kept constant. However, Klein et al. (1990) have demonstrated that a change in line-broadening by a factor of two results in a change in S of about 30 %.

Here, in the Raman scattering experiments of Fig. 5.5, a ratio of the zero-phonon line to the first overtone has been found in the range of 3:1 down to 5:1, resulting in a Huang–Rhys parameter S of around 0.2. This result is in good agreement with Raman spectra published by Klein et al. (1990), Alivisatos et al. (1989), and Rosetti et al. (1983) for CdSe quantum dots both embedded in glasses and organic matrices. In CdSe quantum dots with $1.9\,\mathrm{nm}$ and $4\,\mathrm{nm}$ radius, Klein et al. (1990) observed a ratio of 2.7:1 and 3:1, respectively. Alivisatos et al. (1989) reported for 45 diameter CdSe dots, dispersed in a polystyrene film, the observation of two phonon satellites besides the zero-phonon line at frequencies of 205, 408 and $610\,\mathrm{cm}^{-1}$, with the peak ratios 9:3:1. It is interesting that they could only find the single mode at $205\,\mathrm{cm}^{-1}$ and no more satellites after changing the organic groups for surface derivation. This is a further hint at the influence of interface polarization on the charge distribution in the quantum dot.

For CdS quantum dots, Shiang et al. (1993) published a systematic investigation of the size dependence of resonance Raman scattering. The authors determined the ratio between the zero-phonon line and the first satellite

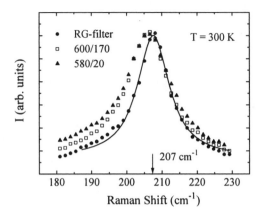

Fig. 5.6. Raman spectra of CdSe quantum dots, grown at 580 °C for 20 h and 670 °C for 170h, showing the asymmetric line shape of the 1LO Raman resonance. From Trinn et al. (1991)

(0ph/1ph) for radii of CdS quantum dots ranging from 0.9 nm to 6.4 nm (see Table 5.1). They studied crystallites that were grown in different environments, but the general tendency of their results was only weakly influenced by the solvent. These measurements in CdS quantum dots show the decrease of the coupling parameter S with decreasing size, just the opposite tendency as supposed for CdSe. The authors propose, that one reason for the discrepancies between the different measurements lies hitherto in the underestimation of the influence of the excited state lifetime (i.e. the broadening Γ) as a crucial parameter. Furthermore, they suggest differences in the charge distributions of the excited states due to different efficiencies for surface localizations in the two materials, CdSe and CdS.

Besides the ratio between the zero-phonon line to its satellites, we can obtain further information about scattering processes from the peak position and line shape of a single resonance in the Raman data. In quantum dots, the optical phonons should have frequencies which are close to the zone-center frequencies of the bulk phonons. In fact, these frequencies have been found in the resonance Raman spectra of Fig. 5.5 with the line center of the zero-phonon peak at 207 cm^{-1}, in comparison with the bulk value of 210 cm^{-1} (Landolt–Börnstein 1982). The presence of the expected SO modes is revealed by the asymmetry of the Raman band with the tail towards lower energies and the small red-shift of the peak energy. In Fig. 5.6, this effect can be seen for the line shapes of the 1LO Raman resonances in relation to the size of the CdSe quantum dots. The existence of contributions from surface modes is indicated by the drop to lower frequencies which is stronger for smaller sizes. Similar experimental evidence for the SO modes, partly more clearly resolved, has been obtained, for example, by Klein et al. (1990). Frequency shifts and line broadening of the phonon modes can also arise from the onset of phonon confinement (Campbell et al. 1986, Tanaka et al. 1992, Rodden et al. 1995, and Chamberlain et al. 1995). The argument of these authors is the

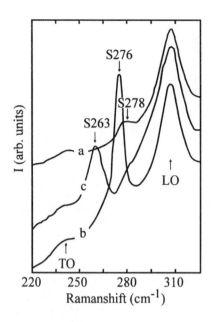

Fig. 5.7. Raman shifts of CdS quantum dots in different matrix materials: (a) polystyrene, (b) polyvinyl alcohol, (c) polymethyl metacrylate. From Pan et al. (1990)

relaxation of the momentum-selection rule in first-order Raman scattering if the size of the nanocrystals decreases. In glassy matrices however, owing to the imperfections at the boundary semiconductor–matrix, in general no distinct peaks below the LO-phonon energies or between the LO and TO energies can be resolved unambiguously and assigned to either SO modes or confined LO modes.

A clear experimental identification of the surface vibration modes has been given by Pan et al. (1990) investigating the Raman spectra of CdS embedded in different organic materials such as polystyrene, polyvinyl alcohol, and polymethyl acrylate (Fig. 5.7). Clear peaks can be resolved shifted by some cm^{-1} compared to the 1LO Raman-resonance line. The assignment to surface modes is based on the analysis of the peak frequencies in relation to different dielectric constants for the host materials. Hayashi and Kanamori (1982) gave the experimental proof of a surface phonon mode appearing between the LO- and TO-phonon for GaP microcrystals. De la Cruz et al. (1995) considered the interface phonons of GaAs quantum dots in $Al_x Ga_{1-x} As$ as a function of alloy composition.

Peaks in the Raman spectra just below the energy of the exciting light (low-frequency Raman scattering) are normally attributed to confined acoustic phonons as dicussed above in the example of elastic vibration of a sphere. This type of acoustic phonons has, among others, been reported by Duval et al. (1982). The authors observed a low-frequency Raman peak around $10\,cm^{-1}$ arising from $R = 5$ to $20\,nm$ microcrystals of spinel embedded in glass. Similar low-energy Raman peaks have been reported to date for silver particles in SiO_2 films (Fujii et al. 1991), for $CdS_x Se_{1-x}$ microcrystals in glass (Champagnon et al. 1993, Roy et al. 1996), and for CdS microcrystals in GeO_2 (Tanaka et al. 1993).

5.1.3 Photoluminescence and Photoluminescence Excitation Spectroscopy

As shown in Sect. 5.1.1, the shifted harmonic oscillator model, i.e., the model where the ground and excited states are shifted in their vibrational coordinates as a result of lattice relaxation, provides in the Huang–Rhys parameter S the information on the coupling of electron–hole pairs and phonons. This parameter can be determined by comparing the difference in the peak maxima of absorption and emission. When the main dephasing process is the coupling of electron–hole pairs to phonons, in the framework of this model correspondence should exist between the results for S from the resonant Raman measurements, from the shift of the luminescence with respect to the absorption, and from the homogeneous linebroadening obtained in nonlinear optics. In this context, the term 'Stokes shift' is originally defined as the difference between the envelope of the absorption and luminescence spectra according to Fig. 5.2. However, in the literature, the term 'Stokes shift' has frequently been introduced to discuss any shift between the absorption and luminescence spectra. In this case the origin of the 'Stokes shift' can be manifold and many more mechanisms are involved. Some examples will be given below.

Figure 5.8 shows typical features of linear absorption and emission spectra measured in II–VI and I–VII quantum dots of different sizes. Usually no size-selective excitation is applied in survey experiments and the emission spectra comprise the radiative recombination from quantum dots within the whole size distribution. An example is given in Fig. 5.8a showing the luminescence spectrum after above-gap excitation at 2.6 eV for small CdSe quantum dots with radii R smaller than the bulk excitonic Bohr radius a_B. The energy separation between the absorption and luminescence maximum exceeds the LO-phonon energy, thus suggesting a strong 'Stokes shift' owing to one-pair–LO-phonon coupling. However, when comparing the spectrum of Fig. 5.8a with that of Fig. 3.16 (size-selective excitation), it can be clearly seen that a shift in the luminescence maximum can already arise from exciting different parts of the size distribution. For quantum dots of larger sizes (weaker confinement), this effect becomes weaker, as indicated in Fig. 5.8b and 5.8c for CdS and CuBr quantum dots. Here, the luminescence is shifted in energy by 20 to 30 meV, a value below that of the LO-phonon energy (the maximum of the luminescence for the CuBr QDs is shown by vertical bars). However, this smaller shift in energy of the luminescence spectrum does not reflect the pure coupling of electron–hole pairs and LO phonons. Evidently, it can be influenced by trapping and capture in defects as observed for the smallest dot radii in Fig. 5.8c. For these small CuBr quantum dots, the luminescence spectrum is independent of size and shows its maximum at constant energy,

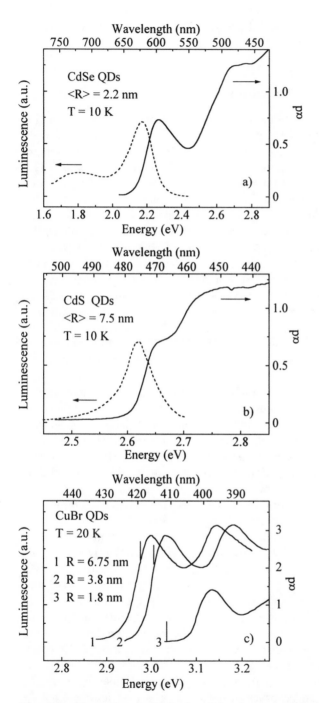

Fig. 5.8. Absorption (solid line) and luminescence (dashed) spectra of (**a**) CdSe quantum dots with $R = 0.4a_B$, (**b**) CdS quantum dots with $R = 2.7a_B$ and (**c**) CuBr quantum dots with $R = 5.4a_B, 3.0a_B$ and $1.44a_B$ (in (**c**) the vertical bars indicate the maximum of the corresponding luminescence spectra). From Woggon (1996a)

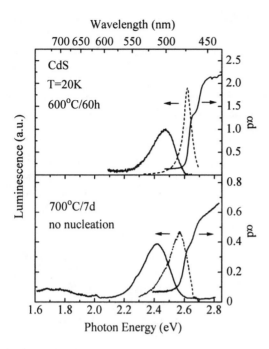

Fig. 5.9. Absorption and luminescence spectra of CdS quantum dots and analysis of the peak shifts in relation to excitation intensities and growth regimes (see text). From Woggon et al. (1994c)

even if the dot radii are further decreased. Thus, components resulting from low-energetically shifted luminescence caused by traps can again lead to an overestimation of the coupling between electron–hole pairs and LO phonons.

A further example of the fact that the red-shift in luminescence is strongly affected by growth parameters and excitation conditions, such as pump intensity and energy, is given in Fig. 5.9. For two representative samples of CdS quantum dots, which were grown using the same matrix composition of 56 % SiO_2, 8 % B_2O_3, 24 % K_2O, 3 % CaO, and 9 % BaO but different annealing procedures, the luminescence spectra were investigated at different excitation densities (Woggon et al. 1994c). The first sample was obtained from a diffusion-controlled process near nucleation, carried out at the low temperature of 600 °C over 60 h (labelled in the following as 600 °C/60 h). The second sample was grown without preceding a nucleation phase directly at high temperatures to promote the coalescence and avoid the normal growth stage (labelled 700 °C/7 d/no nucleation). For the sample 600 °C/60 h, the sizes of the quantum dots were determined by small-angle X-ray scattering to $R = 7.5$ nm with an average deviation of 15 %. For the samples grown at 700 °C, a somewhat larger average radius around 9 nm can be estimated from the absorption onset at lower energies compared with the sample 600 °C/60 h.

The luminescence of both samples was measured under different excitation conditions. The dashed lines in Fig. 5.9 show the luminescence under high excitation using a dye laser tuned resonantly to 2.7 eV. The solid lines show the low-density luminescence spectra of the 600 °C/60 h and the 700 °C/7 d samples excited with a mercury lamp in the UV region. At this low intensity a broad, low-energetically shifted luminescence can be seen, which for sample 700 °C/7 d is accompanied by a second band around 1.7 eV and is not found for sample 600 °C/60 h. The luminescence spectra at different excitation intensities show saturable and nonsaturable parts which allow us to attribute the recombination processes to impurities and direct electron–hole-pair recombination. In sample 600 °C/60 h the trap density is small and its luminescence saturates with increasing pump intensity. The direct electron-hole-pair recombination is characterized by a sharp luminescence band with the maximum 20 meV below the absorption (dashed line). On the other hand, the trap luminescence of the sample grown at 700 °C cannot be saturated and determines the line-width of the luminescence peak even if excited by strong laser intensities.

Calculating the Huang–Rhys parameter S from the luminescence Stokes shift $\Delta_{Stokes} = 20$ meV and from the line-width $\Delta_{FWHM} = 35$ meV for sample 600 °C/60 h, we obtain $S = 0.285$ and $S = 0.25$ respectively, which gives the right order of magnitude ($\hbar\omega_{LO} = 35$ meV for CdS). For the sample 700 °C/7 d, the values of S for the shift and the line-width could not be brought into agreement.

At present, besides the consideration of trap states, a variety of further proposals exists to explain the low-energy shifted emission compared with the absorption. Firstly, the Franck–Condon concept has been extended by considering localization effects. High-quality CdSe quantum dots of the very small size of 2.3 nm diameter have been grown leaving the Se p orbitals uncoordinated at the surface, thus acting as a hole trap (Nirmal et al. 1993, Norris et al. 1993). At low temperatures up to three phonon lines have been observed in the luminescence spectra, the ratios of which undergo dramatic changes if the temperature is raised from 1.75 K to 10 K. The Huang–Rhys parameter S decreases from 3.5 to 1.1 in that temperature range. It has been suggested by Nirmal et al. (1993) and Norris et al. (1993) that the hole is thermally activated and becomes delocalized with increasing temperatures. Owing to the increasing overlap of the electron and hole wavefunction the coupling to the LO phonons diminishes correspondingly.

Secondly, the existence of exciton–phonon complexes and excitonic polarons has been suggested for CuCl quantum dots in the weak confinement range by Itoh et al. (1995). In this paper a red-shifted peak below the excitation is attributed to polarons (confined 1s-exciton bound to a LO phonon) showing increasing binding energy if the size is decreased. A further low-energy shifted luminescence shows strong resonance behavior when the excitation energy equals the energy between the 1p and 1s exciton states plus one or two of the LO-phonon energies. Thus a luminescence line develops as

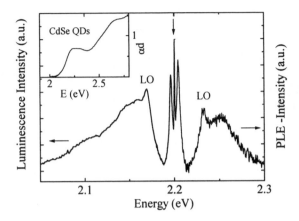

Fig. 5.10. Linear absorption (αd), photoluminescence (PL) and photoluminescence excitation (PLE) of CdSe quantum dots with $\bar{R} = 2.6\,\text{nm}$. The detection / excitation energy 2.2 eV is indicated by the signal of the scattered pump light. From Gindele et al. (1996)

a result of recombination from a 1s-state after excitation of a 1p-state with the help of an LO phonon.

Finally, as already outlined in Chap. 3, for strongly confined II–VI quantum dots the lowest pair state exhibits an internal structure since the degeneracy is lifted because of its nonspherical shape, wurtzite crystal structure and exchange interaction (Efros 1992a; Efros and Rodina 1993; Takagahara 1993a; Nomura et al. 1994; Chamarro et al. 1995, 1996; and Nirmal et al. 1995). This aspect of fine structure might initiate a reconsideration of the interpretation of emission spectra with respect to the 'Stokes shift' and the problem of coupling between electron–hole pairs and phonons. The problem is illustrated in Fig. 5.10 showing the high-resolution emission and excitation spectra of CdSe quantum dots with $\bar{R} = 2.6\,\text{nm}$. The pump energy for the luminescence measurement and the detection energy for the excitation spectrum have been chosen at 2.2 eV in the tail of the linear absorption. Ultranarrow lines caused by the splitting of the lowest pair state can be detected just near the excitation/detection energy and with hardly any 'Stokes shift'. Both in the PL and PLE spectrum the first LO-phonon replica is seen stemming from the optically forbidden transition. In this model the broad band is explained by a superposition of different excited pair states involving the $J = 1/2$ hole state (see Fig. 3.12). The individual pair states obviously experience different types of dephasing (broad bands and narrow lines). Hence, the success in resolving the fine structure simultaneously resulted in new open questions concerning the analysis of the electron–hole pair phonon coupling of CdSe quantum dots with radii between 1.5 and 3 nm in this special situation.

Photoluminescence is likewise capable of displaying the coupling to acoustic phonons. Investigating the time and temperature dependence of the luminescence of CdS quantum dots in the medium confinement, Misawa et al. (1991) obtained experimental evidence for the coupling to low-frequency

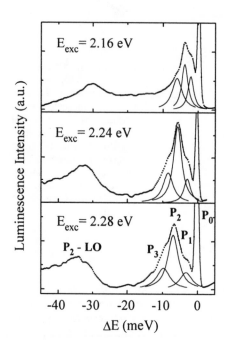

Fig. 5.11. High-resolution photoluminescence at different excitation energies showing the fine structure with size-dependent energies for the peaks P_1, P_2 and P_3 (P_0 corresponds to the pump laser). From Woggon et al. (1996c)

acoustic phonons. A significant quenching of superradiance above a threshold temperature of 45 K has been explained by a dephasing mechanism which involves scattering with confined acoustic phonons. The thermal threshold energy has been related to the energy of a radial compression mode of the CdS sphere.

On investigation of the photoluminescence of the CdSe quantum dots from Fig. 5.10 with higher spectral resolution, a substructure of three peaks has been found within the spectral range attributed so far to the forbidden $(1S_{3/2}, 1s_e)$ pair state with total angular momentum 2 (Fig. 3.12). In Fig. 5.11, the luminescence is plotted for a small spectral range of only 50 meV showing the peaks labelled P_1, P_2 and P_3 a few meV below the pump (P_0). The size dependence can be pursued by tuning the excitation energies between 2.16, 2.24 and 2.28 eV. The individual line-widths of the single peaks are very small and amount to only 1.5 meV FWHM. The single peaks have a different size dependence which has been analyzed in Fig. 5.12 to identify the nature of the three transitions. There the size dependence of the energies of the peaks P_1 to P_3 is plotted. The radii were obtained from a calibration curve after measuring the average radii of a set of samples by different methods and additional comparison with the literature. The absorption process takes place into the allowed $(1S_{3/2}, 1s_e)$ state with total angular momentum 1 (P_0). Peak P_1 and P_3 can be fitted by a $1/R$ dependence, typical for confined acoustic phonons, in contrast to the stronger size dependence of peak

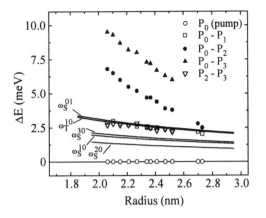

Fig. 5.12. Size dependence for the peaks P_1, P_2 and P_3 and theoretical curves for the eigenfrequencies ω_S^{ln} of the spherical and ω_T^{ln} of the torsional acoustic phonons. From Woggon et al. (1996c)

P_2. This peak shows the dependence $\Delta E \sim 1/R^3$, just the expected enhancement for the short range exchange splitting energy. Therefore, the peak P_2 has been assigned to the split-off, optically forbidden $(1S_{3/2}, 1s_e)$ state. The LO-phonon emission likewise starts from that state, because the energy separation between P_2 and LO has the constant value of ~ 26 meV independent of the pump energy, whereas the energy distance of the LO peak to the excitation pump increases with decreasing radius. The two peaks P_3 and P_1 are attributed to acoustic phonon replica of both the forbidden transition P_2 and of the allowed, strongly absorbing transition P_0 (Woggon et al. 1996c).

The origin of these low-energy acoustic phonons has been attributed to oscillation of an elastic sphere (see Sect. 5.1.1). For CdSe with $c_\ell/c_t = 2.3$ $(c_\ell = 3.63 \times 10^3$ m/s, $c_t = 1.59 \times 10^3$ m/s), the frequencies of the eigenmodes fitting the experimental data are $\omega_T^{10} = 5.8c_t/R$ and $\omega_S^{01} = 6.0c_t/R$, where R is the radius of the sphere. Thus, agreement with the experiment consists for the so-called breathing mode ω_S^{ln} with $l = 0$, $n = 1$, where the whole sphere exhibits dilation and for the torsional mode ω_T^{ln} with $n = 0$ and $l = 1$. At least the spherical mode with $l = 0$ should be able to couple to the electron–hole pair. The modes with $l = 1, 2, 3$, $n = 0$ characterized by large amplitudes at the surface have too small energies to explain the experimental data. By plotting the difference between P_3 and P_2, the nature of the peak P_3 can be attributed to that of peak P_1. Thus Fig. 5.11 shows the discrete confined acoustic phonons as replica in the photoluminescence spectra of quantum dots.

5.1.4 Hole-Burning and Saturation Spectroscopy

A situation similar to that of the analysis of the Stokes shift in luminescence also occurs in part in nonlinear spectroscopy, that is when applying experimental methods which derive the dephasing mechanisms from the homogeneous line broadening. Very often spectral hole burning or laser saturation spectroscopy which we introduced in Chap. 3, have been used to investigate the temperature dependence of the homogeneous line-width $\Gamma(T)$ (and to continue the discussion of the phonon contribution to the dephasing process).

The hole burning data published earlier dealing with CdSe quantum dots in glass matrix showed some common features: for dot radii R between 1.6 nm and 3 nm a large line-width of the spectral holes in the range from 90 meV to 140 meV (FWHM) has been measured and shown to be nearly independent of temperature. This finding would correspond to dephasing times of a few femtoseconds leading to some confusion in search of a suitable fast dephasing process. At that time great divergence existed between the results obtained for quantum dots embedded in glasses and those embedded in organic environment (Roussignol et al. 1989; Peyghambarian et al. 1989; Park et al. 1990; Henneberger et al. 1992; Bawendi et al. 1990; Alivisatos et al. 1988). The results of hole burning experiments are completely different if the same II–VI material is incorporated in an organic matrix. Here organometallic reactions at the surface can be utilized to produce quantum dots, which are passivated or 'capped' by organic groups. Under these conditions a homogeneous line-width of 16 meV has been reported at low temperatures for 1.6 nm radius CdSe-dots (Bawendi et al. 1990; Alivisatos et al. 1988). Obviously, the interface configuration plays an essential role in phase destroying scattering processes in samples showing broad line-width in hole burning experiments (see Sect. 5.3).

The complex spectral behavior of the burnt holes was already mentioned in the hole burning experiments published in 1988 by Alivisatos et al. for strongly confined CdSe quantum dots. Narrow hole-widths could be observed if pumping in the tail of the linear absorption, as well as broad, structured hole burning spectra for excitation above the resonance. In most of the spectra a narrow band is superimposed onto a broad background. For the narrow structure in the hole burning spectra, the temperature dependence has been analyzed and changes in homogeneous line broadening have been measured by a factor of two in the temperature range between 10 and 60 K. The temperature-dependent increase of Γ even at low temperatures again implies the coupling to low-frequency acoustic modes. Owing to the predominance of the cubic crystal lattice in the quantum dots studied, no vanishing of degeneracy and splitting of states is expected. Therefore the spectral shape of the burnt hole has been discussed in the frame of the Franck–Condon model via the narrow zero-phonon line and the phonon replica. The phonon modes are attributed to a radial compression mode of the CdSe quantum dot.

A similar temperature dependence of homogeneous line broadening has also been reported for CdSe quantum dots in glasses when investigating samples with stable interface configurations and very sharp nonlinear resonances ($\Gamma \approx 10$ meV, Woggon et al. 1993b). Between 10 and 40 K a small linear increase in the line-width has been observed with increase in temperature. The value $T = 50\,\mathrm{K}$ can be considered a threshold temperature. Above it a remarkable increase of the line broadening occurs since the interaction of excitons with optical phonons gains increasing importance. The energetic position of the linear absorption peak is nearly constant up to $50\,\mathrm{K}$ and shifts above a temperature of $70\,\mathrm{K}$ up to room temperature with $dE/dT = 3\mathrm{x}10^{-4}\,\mathrm{eV/K}$ to the red, similar to bulk material. These temperature dependencies of both the width of the spectral hole and the energy of the lowest electron–hole-pair transition suggest that coupling to phonons is involved in both cases. The contributions from inhomogeneous broadening and surface effects are considered to be essentially temperature independent. The derivation of the individual line broadening becomes more sophisticated if entering the range of dot radii for which the excited pair states are closer in energy and bleached simultaneously (see the discussion in Sect. 3.1). For a clear identification and investigation of size-dependent coupling mechanisms, a combination of data from different methods is always advisable[2].

The problem of the coupling between electron–hole pairs and phonons is interesting not only for II–VI quantum dots. Similar considerations have also been made for CuCl quantum dots in the weak confinement range (Masumoto et al. 1989a; Wamura et al. 1991). Based on relation (3.30), the saturation intensity $I_{\mathrm{sat}} = \epsilon_0\, c\, \hbar^2\,/\,(2|\mu|^2\, T_1\, T_2)$ can be used to determine the lifetime broadening and, by this, the absolute value of the homogeneous broadening (Masumoto et al. 1989a). For the Z_3 exciton in CuCl quantum dots with $R = 6.1\,\mathrm{nm}$, the value of $\Gamma = \hbar/T_2$ has been determined to be $0.18\,\mathrm{meV}$. Compared with II–VI semiconductor dots, this value is extremely small. By calculating the transition dipole moment from the data of laser saturation spectroscopy, an enhancement by a factor of 13 is obtained in comparison with the bulk value of CuCl. The temperature dependence of Γ for weakly confined CuCl quantum dots is measured for the range between $20\,\mathrm{K}$ and $300\,\mathrm{K}$ (Masumoto et al. 1989a) and 2 and $60\,\mathrm{K}$ (Wamura et al. 1991) by spectral hole burning. In the low-temperature range the dependence $\Gamma \sim \Gamma_0 + CT^2$ has been found and explained by acoustic phonons. In the range above $20\,\mathrm{K}$, the line broadening is fitted best by $\Gamma(T) = \Gamma_{\mathrm{inh}} + a/[\exp(B/k_{\mathrm{B}}) - 1]$. The latter relation has been ascribed to the interaction of Z_3 excitons with LO phonons.

Applying the technique of persistent spectral hole burning, the observation of confined acoustic phonons succeeded in the spectra of CuCl quantum dots embedded in glass and in NaCl (Okamoto and Masumoto 1995). The phonon side bands have been resolved and the observed energies coincide with the ones calculated for the vibration of an elastic CuCl sphere.

5.1.5 Four-Wave Mixing

After discussing methods investigating the nonlinear absorption, we now turn to experiments predominantly exploiting the nonlinearity in the refraction

---

[2] However, at high densities, spectral hole burning in the gain spectrum works very well to identify the dephasing times at high carrier densities.

index to study dephasing processes. In the following, results will be reported which have been obtained by applying the widely used different types of four-wave mixing experiments. Recall that in degenerate four-wave mixing (DFWM), two laser beams of frequencies $\omega_1 = \omega_2$ produce a spatially periodic intensity inside the sample attended by a periodic modulation in polarization and population and, thus, also in the refraction index and absorption coefficient. If light passes the sample a diffraction pattern is found with a diffraction efficiency proportional to the contrast of the density modulation and polarization. This provides insight into, for example, the mechanisms of nonlinearity, or the diffusion constants, or the lifetime and dephasing time of the excited carriers. Time resolution is achieved when delaying one laser beam with respect to the other. Four-wave mixing experiments can be carried out in a variety of configurations and applying more than two laser beams (see e.g. Eichler et al. 1986).

The majority of published dephasing studies in quantum dots concern the direct determination of the T_2 time itself (e.g. Nuss et al. 1986). Mostly $CdS_{1-x}Se_x$ quantum dots in commercially available glasses have been investigated. The values of the T_2 times gained lies in the range of some tens of femtoseconds. These very short dephasing times imply the occurrence and superposition of more than one dephasing mechanism. Over long periods of time, the distinction between the different contributions to the dephasing was complicated by the lack of femtosecond laser sources with large spectral tuning range and high intensity. Therefore, systematic investigations of T_2 in dependence on temperature, intensity and over a great range of dot sizes are rare and one of the most important tasks for future research. Here, with the tremendous development of femtosecond laser techniques, we can expect new impulses in the field of the ultrafast spectroscopy of quantum dots.

By use of a three-pulse photon echo technique, Schoenlein et al. (1993) succeeded in measuring the pure electronic dephasing time for $R = 1.1$ nm CdSe quantum dots as $T_2 = 85$ fs at 15 K. Based on a colliding-pulse-mode-locked (CPM) laser system, short pulses of ~ 20 fs were produced and amplified within a spectral range of 2 eV to 2.6 eV. This set-up has made possible a detailed study of the size-dependent dephasing in the case of small CdSe quantum dots with radii between $R = 1$ and 2 nm (Mittleman et al. 1994; Schoenlein et al. 1993). The authors divide the experimentally observed scattering processes into three types: (i) nonpolar interaction of electron-hole pairs with a heat bath of low-frequency acoustic phonons (heat bath coupling), (ii) phase loss due to fast relaxation of the electron–hole pair into a long-living state, for example, a trap state at the interface (lifetime broadening), and (iii) elastic scattering at imperfections in the quantum dot boundary.

To separate these different contributions experimentally, a three-pulse configuration is used where the third pulse has been delayed by exactly the LO-phonon vibrational period in order to suppress the modulation of the photon echo signal by these coherently excited LO phonons. As a result, the

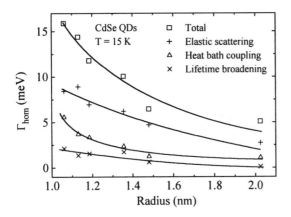

Fig. 5.13. Leading contributions to the homogeneous line broadening of strongly confined CdSe quantum dots measured by a femtosecond three-pulse photon echo technique by Mittleman et al. (1994)

size dependence of the contributions (i) to (iii) has been determined as shown in Fig. 5.13. For the coupling to the heat bath of confined acoustic phonons, a $\sim 1/R^5$ relation has been found reflected in the size-dependence of the homogeneous line broadening Γ and explained by a deformation potential coupling scheme. For the small CdSe quantum dots investigated, the line broadening due to low-frequency phonons shows a linear dependence on temperature and represents a significant contribution to the homogeneous line-width of the optical transitions up to room temperature. These results are in good agreement with theoretical calculations of Takagahara (1993b) discussed in Sect. 5.1.1 (see Fig. 5.4). For the lifetime broadening and the elastic scattering, a $\sim 1/R$ dependence yielded the best fit for the size dependence of Γ. Thus, the size-dependence of contribution (ii) and (iii) is determined by the change in the surface/volume ratio of the dot. When studying the coupling strength of the electron–hole pair to the LO-phonon mode, no monotonous size-dependence could be derived and values around 0.6 have been obtained for the Huang–Rhys parameter S. This finding has been explained by modifications in the dipolar character of the wave function, again demonstrating that knowledge of the exact symmetry of the pair-wave function is a prerequisite for understanding the polar coupling strength in quantum dots.

A further experimental attempt to study the phase relaxation in II–VI quantum dots is the use of four-wave mixing experiments with incoherent light (Huang and Kwok 1992; Acioli et al. 1990a; Tokizaki et al. 1989; Misawa et al. 1988). These experiments were performed on commercially available $CdS_{1-x}Se_x$-doped glasses and experimental glasses doped with very small CdS and CdSe quantum dots. Two laser beams with small coherence length form a grating in the sample. One of the lasers is delayed in time with respect to the other. If the dephasing time T_2 of the sample is large, then the grating still exists over long delay times due to the phase preservation in the sample. The T_2 time can be deduced from the maximum delay of the two lasers,

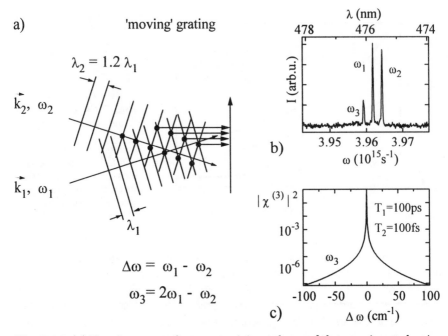

Fig. 5.14. (a) Non-degenerate four-wave mixing: scheme of the experiment showing the formation of the moving grating by interference of two lasers with different frequencies ω_1 and ω_2, (b) Signals detected by the OMA system in the NDFWM setup: ω_1 and ω_2 are the scattered light of the two exciting laser pulses, and ω_3 is the frequency-shifted diffracted signal. The measured sample is a CdS-doped glass with 7.5 nm quantum dots. (c) Intensity of the first diffracted order at ω_3 as a function of detuning $\Delta\omega = \omega_1 - \omega_2$ calculated for the dephasing times $T_2 = 100$ fs, and a lifetime $T_1 = 100$ ps. From Woggon et al. (1995c)

which still give a detectable diffraction signal. However, again the T_2 times obtained in this type of FWM are very short – between one and some tens of femtoseconds. No systematic size or temperature dependence has been reported.

Experimental methods providing femtosecond time-resolution but working in the frequency domain have the advantage that they are based on ns laser systems with spectrally narrow excitation pulses and large tunability. Among these methods, nondegenerate four-wave mixing will be introduced in more detail. In the nondegenerate case of FWM, the frequencies of the two laser beams are different, $\omega_1 \neq \omega_2$. If one of the pump beam is detuned by a small difference $\Delta\omega$ compared to the other, then the interference results in a 'moving grating' structure and produces a nonstationary intensity distribution. The third-order nonlinearity of the material produces a signal at $\omega_3 = 2\omega_1 - \omega_2$. When the two incident frequencies are both resonant with the inhomogeneously broadened optical transition and $\Delta\omega = |\omega_1 - \omega_2|$ is in the range of the homogeneous line-width, NDFWM can serve for the determination of the dephasing time. The scheme of the experiment is shown in Fig. 5.14 illustrating the formation of the moving grating by interference of two lasers with different frequencies, ω_1 and ω_2.

The theoretical description of the NDFWM for resonant excitation of an inhomogeneously broadened optical transition has been developed by Yajima and Souma (1978) and Yajima et al. (1978). The simplest models to introduce the phase relaxation are the optical Bloch equations formulated for a two-level system. Although the correct understanding of an optically excited bulk semiconductor requires a many-body formalism, the simple two-level system should at least qualitatively provide a description of the relaxation dynamics in quantum dots with their discrete level structure. Starting directly from the equation of motion for the density matrix, ρ has been calculated in perturbation theory by an expansion in a power series of different orders of perturbation (Yajima and Souma 1978; Yajima et al. 1978). For the output light intensity I at $\omega_3 = 2\omega_1 - \omega_2$ holds $I \sim |\chi^{(3)}|^2$ with

$$|\chi^{(3)}|^2 = \frac{|K_3|^2}{(1 + \Delta\omega^2 T_1^2)(1 + \Delta\omega^2 T_2^2)} \tag{5.20}$$

where K_3 is

$$K_3 = -2\pi |d_{21}|^4 n\rho_0 S(\Omega) T_1 T_2 \hbar^{-3} \tag{5.21}$$

and d_{21} the dipole matrix element, n the density of two-level atoms, and $S(\Omega)$ the size distribution function normalized in energy. This result is illustrated in Fig. 5.14c with reasonable parameters of T_1 and T_2. The dependence around $\Delta\omega = 0$ is caused by the value of the lifetime T_1. A sharp peak is expected if T_1 is large compared to T_2. The wings in the detuning curves are determined by the dephasing time T_2. The ratio of the signal intensity at zero detuning compared to the signal intensity for large values of $\Delta\omega$ is determined by the ratio between the times T_1 and T_2. In the corresponding experiment, the strength of the decrease of the diffraction efficiency with increasing detuning is a measure for the dephasing time T_2. NDFWM is a useful tool for investigating isolated quantum dots in glasses because the diffusion processes are of minor importance: the expected T_2 times lie in the range below 1 ps and are small compared to the lifetime T_1.

The temperature dependence of the dephasing time has been investigated in CdS quantum dots of 7.5 nm radius at an excitation energy that corresponds to the lowest transition between the quantum-confined electron–hole-pair states. The NDFWM experiment has been performed at an intermediate intensity and by adjusting the laser frequencies for $\Delta\omega = 0$ according to the temperature dependent shift of the linear absorption spectrum. Figure 5.15 shows the intensity of the first diffracted order at ω_3 versus the detuning of the two laser frequencies for three different temperatures of 8, 50 and 300 K. The peak around $\Delta\omega = 0$ represents the lifetime T_1 of the underlying population grating (values of T_1 larger than 20 ps cannot be completely resolved with this experimental setup, which is limited by the finite line-width of the

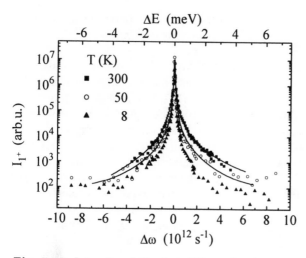

Fig. 5.15. Intensity of the first diffracted order versus detuning of the two laser beams with the temperature as parameter, for 7.5 nm CdS quantum dots. From Woggon et al. (1995c)

exciting laser beams and the spectral resolution provided by the diode array behind the spectrometer). From a fitting process of the detuning curves according to (5.20) and (5.21) a faster dephasing, i.e., a decrease in T_2, has been found with increasing temperatures. The corresponding data for the results of Fig. 5.15 are $T_2 = 400$ fs (8 K), 300 fs (50 K) and 180 fs (300 K).

This change in T_2 with temperature is significant, but not dramatic. Between 8 K and 50 K, the result is similar to that reported in Woggon et al. (1993b) and Mittleman et al. (1994) for strongly confined CdSe quantum dots using both spectral hole burning and the three-pulse photon echo technique described. There a decrease in the dephasing time by a factor of 0.75 has been found in the range between 10 and 50 K for dot radii ranging from 1 to 2 nm. Compared with the results obtained for small CdSe quantum dots, the temperature-dependent dephasing of the CdS QDs investigated here is considerably smaller, in particular at higher temperatures from 50 K up to room temperature. This experimental finding can be understood assuming a smaller charge separation of the electron and hole wave functions in larger dot sizes, and therefore, a smaller coupling of the excited electron–hole pairs to LO phonons. Likewise, the coupling to low-frequency acoustic phonons is reduced compared with strongly confined quantum dots.

5.2 Energy Relaxation

One of the most intensively discussed questions is the relation between the time scales of phase and energy relaxation. Strictly speaking, any energy relaxation can be simultaneously considered a mechanism of phase loss. At least two problems have been proven to be essential in the course of this discussion. The first one concerns the energy separation between the confined levels of the quantum dots, the second consists in the number of states acting as free final states for scattering processes, which is drastically diminished in quantum confined structures. In addition, confinement-induced valence-band mixing couples the wave functions from different excited hole states.

In high-quality samples and at low excitation densities the relaxation through the ladder of excited states can only proceed via multiphonon processes. When increasing the sizes of the quantum dots, the spacing between the confined energy levels increases and the energy distance can become larger than the phonon energies. A considerable modification of the electron and hole relaxation process might occur because of the lack of lower/higher energy states of electrons and holes matched in energy for the phonon absorption and emission. This consideration has led to speculations about the suppression of the relaxation rate within the ladder of the confined energy levels and the existence of a so-called 'phonon bottleneck'. Benisty et al. (1991) concluded that the suppressed excited carrier relaxation presents an intrinsic mechanism for the poor luminescence of etched quantum box structures.

During the last years, the intense theoretical work dealing with the problem of phonon bottleneck has been extended to quantum dots (Bockelmann and Bastard 1990; Benisty et al. 1991; Nojima 1992; Bockelmann 1993; Benisty 1995; Efros et al. 1995). The transfer of energy between the electron and hole subsystem and Auger-like processes have been investigated (Bockelmann and Egeler 1992; Efros et al. 1995) as well as the efficiency of multiphonon processes or of the combined LO- and LA-phonon emission (Inoshita and Sakaki 1992). The calculations have not been confined to the treatment of the electron and hole subsystem separately but also Coulomb interaction and exciton relaxation has been involved (Efros et al. 1995; Bockelmann 1993). Generally, the theoretical results show the tendency that interaction with the known, regular acoustic phonons can not remove the problem of the phonon bottleneck (Bockelmann 1993; Bockelmann and Bastard 1990; Nojima 1992; Benisty 1995; Benisty et al. 1991). A window to picosecond relaxation times opens for Auger-like processes where the energy relaxation is mediated by an electron–hole plasma (Bockelmann and Egeler 1992) or via energy transfer from the electron to the hole and fast relaxation within the hole subsystem (Efros et al. 1995). An enhancement of the relaxation rates has also been predicted when considering the interaction with combinations of specific phonons or with new types of confined acoustic vibration modes (Inoshita and Sakaki 1992; Takagahara 1993a). The coupling of the quantum dot levels to a deep trap could likewise provide a rapid relaxation path via multiphonon-assisted tunnelling as proposed by Sercel (1995).

However, a number of experiments have been published which do not indicate the appearance of a phonon bottleneck or of reduced relaxation rates.

For example, in a femtosecond differential absorption measurement, the relaxation behavior has been investigated for CdSe quantum dot samples for the two cases of above-resonant excitation with energies (i) matching the energy level separation by multiples of the LO-phonon energies and (ii) disagreeing with these energy separations. It can be shown that no difference exists for the cases (i) and (ii) and a very fast relaxation dominates with times of a few hundreds of femtoseconds.

We start the discussion with the investigation of the relaxation behavior of CdSe quantum dots with an average radius of $\bar{R} = 2.2$ nm (Woggon et al. 1996d). In Fig. 5.16, the femtosecond differential absorption is shown in the case of very low excitation. In Fig. 5.16a the pump is tuned to an energy in the absorption tail (see insert) and we predominantly excite nanocrystals with sizes larger than the average radius $\bar{R}$. The very small bleaching signal below the pump energy indicates that we are investigating predominantly one size of quantum dots and excite resonantly the energetically lowest one-pair state without any possibility of relaxation. The strong peak around the pump corresponds to the bleaching of the $(1S_{3/2}, 1s_e)$ state and the signal at energies above the pump shows the contribution of (for that size) near-neighboring one-pair transitions composed of the same electron state and excited hole states. The population of the lowest-pair state, visible by the spectral hole created around the pump energy, shows only minor changes in the time range between 100 fs and 1 ps. Remarkable is the fast evolution of the induced absorption feature on the high energy side. Already after 160 fs, almost instantaneously with the pump pulse, the absorption process of a probe photon and the formation of a two-pair state can be observed and after 400 fs two pronounced maxima can be distinguished. Later, the whole differential absorption spectrum is formed by a combination of bleaching of the one-pair states and induced absorption into two-pair states. In the time range > 1 ps, the spectral shape remains constant and the further decay is now due to the beginning of recombination processes. Owing to the careful chirp correction, we can exclude any energy shift of the peaks with delay time.

Figure 5.16a shows the nearly instantaneous formation of the two-pair states simultaneous with the population of the one-pair states. The ultrafast formation of the two-pair states can be used as an additional probe for the relaxation process. The relaxation time to the ground state can be derived by measuring the time until the induced absorption feature at higher energies appears in the spectrum that belongs to the probe-photon absorption process out of the one-pair ground state (see the level schemes discussed in Chap. 3).

In the next experiment the relaxation behavior has been studied for the pump energy tuned to 2.205 eV (562 nm). For an ensemble of dots with $\bar{R} = 2.2$ nm we excite here the quantum dots at the upper end of the size distribution. From Fig. 3.11 it can be seen that for a certain size, selected by the pump energy, the energy separation between the ground and first excited

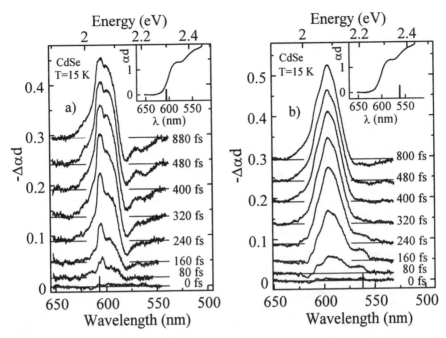

Fig. 5.16. Femtosecond low-density differential absorption spectra of CdSe quantum dots measured at $T = 15$ K and different delay times of the probe beam. The pump beam (vertical bar) is tuned (**a**) to the energetically lowest one-pair transition and (**b**) to energies above the (s,d)-type excited one-pair transitions. From Woggon et al. (1995d,1996a,d)

one-pair state corresponds to cases of $\Delta E_1 = E_0 + N\hbar\omega_{LO}$, i.e., the excess energy matches both energy separation between the $(2S_{3/2}, 1s_e)$ state and the $(1S_{3/2}, 1s_e)$ state and multiples of the LO-phonon energy. An excitation energy of 2.205 eV here is equivalent to $3\hbar\omega_{LO}$.

In Fig. 5.16b, a distinct hole around the pump energy can be observed at early time delay indicating the presence of a nonequilibrium carrier distribution. The spectral hole remains observable a few 100 fs after the pump pulse had passed the sample. However, after 480 fs we detect the induced absorption which belongs to the probe-photon absorption process out of the one-pair ground state. At that time, the hole which is created by the pump at an energy of an excited state is relaxed to its ground state and the probe photon will be absorbed nearly instantaneously causing the induced absorption. From the experiment presented in Fig. 5.16b the relaxation time has been derived to roughly about 0.5 ps.

An interesting case is the following, shown in Fig. 5.17. Now we tune the pump to an energy slightly below that of the experiment shown in Fig. 5.16b. The energy $E = 2.112$ eV (587 nm) does not correspond to any of the multiples of the LO-phonon energy. The bleaching band itself is very broad and

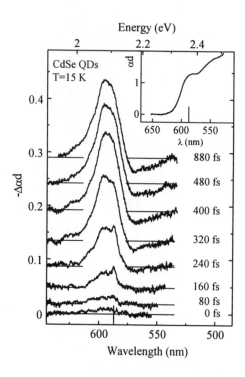

Fig. 5.17. Femtosecond spectral hole burning of CdSe quantum dots for pump detuning off-resonant to multiples of the LO-phonon energies. From Woggon (1996a)

unstructured, but we see that also in the case of off-resonant excess energy, the relaxation behavior is not very different from that of Fig. 5.16. Simultaneously with the disappearance of the spectral hole around the pump, the induced absorption builds up in the time range of ∼0.5 ps.

To conclude, from the experiments presented no hint at the existence of a so-called phonon bottleneck has been obtained, i.e., the suppression of the relaxation rate within the ladder of the confined energy levels because of the lack of LO-phonons matching the level separation. The fast energy relaxation within 500 fs is independent of the relation between level spacing and LO-phonon energy and shows no bottleneck effect. If a so-called phonon bottleneck were significant, much longer relaxation times in the nanosecond range would be expected. Obviously the occurrence of a sufficient number of new, low-energy acoustic phonons introduced by the spherical confinement removes the problem of the phonon bottleneck. Additionally, with increasing intensity the formation of two-pair states (biexcitons) of various energies dominates the population dynamics due to the strong influence of Coulomb interaction in quantum dots even in the strong confinement. At high pair densities, the created two-pair states may create a new relaxation path due to their stimulated decay into one photon and one *relaxed* electron–hole pair with lower energy (see Chap. 3.2.3 and below).

This relaxation dynamics will now be compared with the results for bulk-like $CdS_{0.7}Se_{0.3}$ quantum dots with radii between 18 nm and 30 nm (Fig.

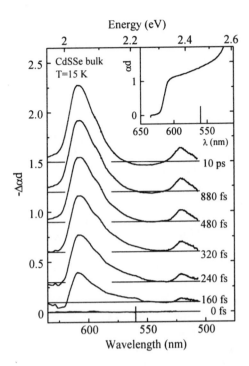

Fig. 5.18. Femtosecond pump- and probe spectra measured at bulk-like $CdS_{0.7}Se_{0.3}$ quantum dots for different probe delay times at $T = 10\,K$ and excitation at 556 nm (vertical bar). From Woggon (1996d)

5.18). A nonequilibrium distribution has been excited by a 100 fs pump pulse at 2.23 eV (556 nm), which is indicated by the burnt hole around the pump energy at early probe delay times. The relaxation time is derived from the lifetime of that burnt hole, since there the change of the carrier distribution is explicitly reflected in the temporal development of the absorption bleaching. As shown in Fig. 5.18, the population around the pump energy lives for about 0.5 ps after the pump pulse. During that time the continuous decrease of the bleaching signal around the pump and the increase of the signal in the tail indicates energy relaxation to the ground state. The bleaching at 2.4 eV is caused by the blocking of the hole absorption in the C valence band owing to population of the confined s-electron in the conduction band. The temporal development of the bleaching spectrum resembles very much the band filling effect known from bulk semiconductors. Similar observations have been reported in several other studies of the relaxation dynamics for larger CdSe quantum dots with $R > a_B$ (Tokizaki et al. 1992; Klimov et al. 1995; Hunsche et al. 1996).

We shall now compare the behavior obtained for the bulk-like $CdS_{0.7}Se_{0.3}$ quantum dots with studies by Fluegel et al. (1992) of the carrier relaxation in bulk CdSe using a similar femtosecond pump-and-probe technique. There, by excitation with a 70 fs pump pulse several LO-phonon energies above the A,B-exciton resonances, a nonthermal carrier distribution was created at the energy of the pump pulse. The burnt transient spectral hole disappeared in less than 100 fs. The rapid decay of the hole has been attributed to an efficient carrier–carrier scattering accompanied by carrier–LO-phonon induced thermalization. The theoretical analysis of the carrier–carrier scattering shows the importance of Coulomb screening for the scattering within a many-particle system (Binder et al. 1992). In particular, bleaching of the A,B excitons is already observed at probe delays just after the pump pulse at times, where the bleaching of the population around the pump energy still dominates the spectrum. A similar situation has been reported by Shah et al. (1987) investigating the hot carrier relaxation in bulk GaAs at high carrier densities.

In analogy to these studies, the carrier relaxation in bulk-like $CdS_{0.7}Se_{0.3}$ is explained by carrier–carrier and carrier–LO-phonon scattering, similar to the bulk. Since the carrier–carrier scattering time is intensity dependent, the slightly longer relaxation times [100 fs in bulk CdSe (Fluegel et al. 1992) compared with 500 fs in Fig. 5.18] can be caused by the smaller pump intensities used to excite the $CdS_{0.7}Se_{0.3}$ quantum dots.

Dealing with that problem by NDFWM, we also found first hints that the dephasing process is different if starting from the first or from the second excited state. When analyzing the detuning curves for $R = 7.5$ nm CdS quantum dots, a faster dephasing time was found for the higher energy states ($T_2 = 300$ fs for $E_p = 2.587$ eV and $T_2 = 250$ fs for $E_p = 2.680$ eV). It seems very unlikely that scattering at interfaces is different for ground and excited states. Therefore the upper state must have an additional relaxation process compared with the lower state, which is just the transition to that state.

The problem of relaxation time can become important for application concepts. Quantum dots in the strong confinement range are believed to act as an ideal gain medium due to the concentration of density of states in a few confined levels. In the context of that model, the stimulation of the one-pair transitions should be easily attainable at very low pump power. For the fast development of an inversion within the ladder of the confined dot levels, the relaxation process becomes crucial.

Figure 5.19 shows the evolution of the gain spectrum with time for CdSe quantum dots of an average radius $\bar{R} \sim 2.5$ nm measured in a fs-pump- and probe experiment (pulse width 115 fs, pump fluence 25 mJ/cm^2, pump energy 2.213 eV) (Giessen et al. 1995, 1996a). The occurrence of optical gain is indicated when the absorption signal αd becomes negative. At early times (320 fs) the gain develops from the low energy side of the spectrum, reaches its maximum after 2 ps and lives then for about 200 ps. The gain region stretches from below the lowest quantum confined absorption transition up to the pump energy. The observation of the spectrally broad gain spectrum both in intensity- and time-dependent experiments (see Sect. 3.2 for comparison) verifies the multilevel feature of the gain in zero-dimensional systems. The extremely fast gain buildup within a few picoseconds is a further confirmation of the absence of the phonon bottleneck. Additionally, there is an relaxation mechanism which can be understood as a pair–pair scattering (interaction)

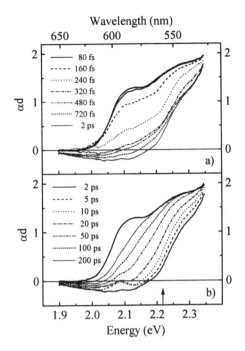

Fig. 5.19. Gain buildup (**a**) and decay (**b**) measured at $T = 10\,\text{K}$ for $R \sim 2.5\,\text{nm}$ CdSe quantum dots. The gain is at its maximum after 2 ps and characterized by a broad quasicontinuous spectrum which extends around and below the absorption edge. It has a lifetime of about 200 ps. The thick line is the linear absorption and the pump energy is indicated by an arrow. From Giessen et al. (1995, 1996b)

with the result of a stimulated emission of a photon and a relaxation of one of the pairs. To what extent Coulomb interaction additionally influences the scattering between the excited electrons and holes is an open question and even more attention should be paid to scattering processes mediated by the Coulomb interaction in quantum dots.

5.3 Scattering at Defects and Interfaces

In the previous sections different examples have shown that the interface is involved in the electronic properties in manifold ways. The interface problem continually reappears through all research on quantum dots. First, in most of the experiments one has to clarify and eliminate its influence before investigating the essential subject.

Strictly speaking, the interface influence cannot be eliminated if the interface localizes one carrier in such a way that an enhancement of charge separation occurs. This property is the base of new phenomena beginning with the described change in the coupling between electron–hole pairs and phonons, the dephasing by scattering at interfaces, or finally the occurrence of real trap processes connected with the red-shift in luminescence owing to the increase in binding energy. These phenomena are closely related. In fact, it is impossible to divide them physically or by language, which often gives rise to some confusion.

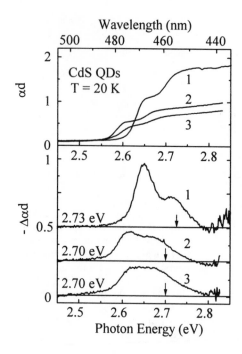

Fig. 5.20. Differential absorption spectra of CdS quantum dots, grown at different conditions, for explanation see text. From Woggon et al. (1994c)

For the sake of completeness, in this section we show the direct comparison of the line-width Γ for three intentionally differently grown CdS-doped samples. We summarize how greatly Γ depends on the choice of the growth procedure. Based on such an analysis, one can choose high-quality samples for the experiments, which increase the chances of the suppression of interface influence.

Figure 5.20 shows the linear and nonlinear absorption spectra of CdS quantum dots arising from different growth stages (Woggon et al. 1994c). We compare the hole burning spectra of two samples of CdS quantum dots which have been used for NDFWM experiments and which are characterized by the growth parameter 600 °C/60 h (1) and 700 °C/7 d/no nucleation (3). Additionally, we show a third sample with parameters similar to (3), namely 700 °C/8 d, but carrying out an additional preannealing step to promote nucleation. This was performed at 560 °C over some days (2).

The differences in the linear absorption spectra (upper part of Fig. 5.20) of the three samples are essentially confined to the different scale in αd caused by the different filling factors. In contrast to small CdSe quantum dots no sharp peaks appear in the larger CdS dots, but two pronounced steps characterize the spectral behavior of the linear spectra. The weak confinement gives only small energy separations between the quantum confined states and the structures merge into shoulders. The step corresponds to the splitting between the A,B and the C valence bands.

The nonlinear absorption ($\Delta \alpha d$) has been measured by nanosecond pump-and-probe spectroscopy with different energies of the pump laser ranging from 2.6 to 2.75 eV. To compare the absorption changes of the different samples in Fig. 5.20, the pump intensity was normalized to excite nearly equal carrier densities. The energy shift of the resonances is small compared to their homogeneous line-width and the homogeneous and inhomogeneous broadening are of the same order of magnitude. Changing the pump energy, the bleaching spectrum is spectrally constant and does not shift with the laser excitation. The width of the bleaching spectrum is strongly dependent on the growth procedure. An estimate of the homogeneous line-width can be made assuming the superposition of Lorentzian resonances within the burnt hole. The full width at half of the maximum (FWHM) is then approximately 35 meV for the first electron–hole pair transition of sample 600 °C/60 h, and 54 meV and 58 meV for the samples 700 °C/8 d and 7 d, respectively ($T = 20$ K). The differences in the line-width of the bleaching spectra of Fig. 5.20 are also preserved when decreasing the pump intensity; thus power broadening is negligible.

We attribute this difference primarily to a dephasing process introduced by scattering processes influenced by the interface properties. Apparently, the different growth procedures produced different interface configurations and thus gave rise to a change in the scattering mechanisms. This holds both for elastic scattering resulting from interface roughness and for changes in the polar coupling mechanisms owing to changes in the polarization of the interface region, modifying the charge distribution of the pair wave function inside the quantum dot.

The absolute value of $\Delta \alpha d$ is significantly larger for the sample 600 °C/ 60 h. For the matrix composition considered, the diffusion controlled growth process near nucleation, carried out at low temperatures, resulted in a saturation intensity approximately one order of magnitude smaller and in a higher nonlinear optical response compared with a high-temperature heat treatment (see also Chap. 6).

5.4 Carrier–Carrier Scattering

In bulk semiconductors the problem of carrier–carrier scattering gains more and more importance when gradually increasing the electron and hole densities. In contrast to three-dimensionally confined systems, bulk semiconductors possess a high number of k states which can act as initial or final states for scattering events. In quantum dots however, the density of available states drastically changes when going from the case of strong confinement to that of weak confinement. Information about the efficiency of carrier–carrier scattering and collisional broadening in three-dimensionally, but weakly confined semiconductors is presently not available from theory, neither for the interaction of electron–hole pairs with additional free carriers nor for other inco-

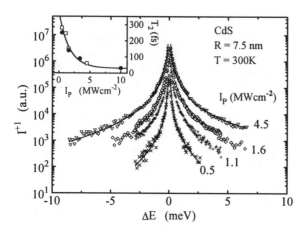

Fig. 5.21. NDFWM diffraction efficiency measured for CdS quantum dots of R = 7.5 nm. The insert shows the dephasing times obtained from (5.20) as a function of excitation intensity. From Woggon et al. (1995c)

herent pairs. For the case of strongly confined semiconductors with a limited number of pair states all populated at high excitation, scattering between these pairs seems unlikely because of missing free final states. In this case a more or less constant dephasing time is expected, independent of excitation intensity.

In the following, examples for both cases will be given. The dephasing has been determined by NDFWM for CdS quantum dots with $R > a_B$ as a function of intensity, and the T_2-time has been studied in the inverted system of strongly confined CdSe quantum dots by spectral hole burning in the gain spectrum.

The influence of high carrier densities on the dephasing time T_2 has been studied in CdS quantum dots by NDFWM with the aim of finding (i) the carrier density, where the interparticle interaction starts to be the dominant process and (ii) the lowest limit of the time T_2 when increasing excitation intensities. It is expected to be in the range of some femtoseconds (Woggon et al. 1995c). In these larger quantum dots, more than one electron–hole pair can be excited, which leads us to expect a strong influence of carrier–carrier interaction on the dephasing.

The results are shown in Fig. 5.21. At low excitation intensities of $500\,\mathrm{kW/cm^2}$, corresponding to about one electron–hole pair per dot, the dephasing time has been determined to be $T_2 = 280\,\mathrm{fs}$. This is one of the largest values reported up to now for the dephasing time in CdS quantum dots at room temperature and has been ascribed to the very low excitation densities applied in the NDFWM measurements as well as the high interface quality of the sample. In this range of excitation the dephasing is mainly attributed to the interaction of the excited electron–hole pairs with phonons (see above). For excitation intensities well above $1\,\mathrm{MW/cm^2}$, i.e., for excitation of a few more electron–hole pairs inside one dot, a significant increase in the wings of the detuning curves can be seen. The shortest dephasing time proven in this

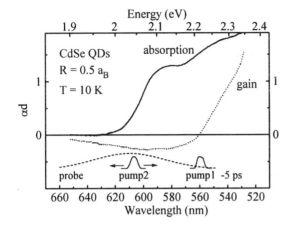

Fig. 5.22. Scheme of a spectral hole burning experiment in the gain spectrum of CdSe quantum dots. Pump pulse 1 arrives at the samples 5 ps before pump pulse 2 and creates the gain spectrum. The usual hole burning experiment then has been performed by tuning pulse 2 and testing the change in the gain spectrum by the probe continuum

measurement is 30 fs, derived from the limit at highest excitation densities. Notice, that the T_2 time when applying the fitting procedure of (5.20), is now, strictly speaking, only a phenomenological parameter, since the underlying model does not involve a many-particle interaction. However, the clear increase of the diffracted signal at ω_3 for larger detuning values $\Delta\omega$ indicates unambiguously the behavior discussed. The insert of Fig. 5.21 shows the dephasing times as a function of excitation intensity both for the transition from the first (open squares) and from the second excited (filled circles) pair state to the ground state. Obviously, when exciting more than one electron–hole pair per dot, which can easily be attained by excitation densities usually applied in experiments of nonlinear optics, the carrier–carrier interaction is the dominant scattering process. The very fast dephasing at high excitation is a clear hint that many-particle interaction becomes an efficient channel of phase loss. Quantum dots of larger sizes therefore show similar behavior as known from the bulk material at comparable excitation densities (Bigot et al. 1990; Fluegel et al. 1992; Binder et al. 1992).

While scattering between the different type of carriers (electrons, holes and excitons) has been widely investigated in two-dimensional structures, only little is known for quantum dots. At present, the direct comparison of the efficiency of collisional broadening caused by the interaction with additional free carriers on the one hand, or with incoherent excitons on the other hand, is not available from the theory of three-dimensionally confined semiconductors. From 2D-structures it is known that excitons are much more efficient in their interaction with free carriers than with other excitons. This fact is explained by the considerably weaker screening of the Coulomb interaction in 2D compared with 3D (Schmitt-Rink 1985). In quantum wires the reduced exciton–exciton scattering has recently been observed by Oestreich et al. (1993).

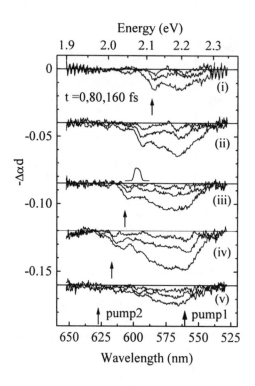

Fig. 5.23. Spectral hole burning in the gain region measured at early delay times of 0 fs, 80 fs and 160 fs with the pump 1 at 560 nm and tuning the pump 2 between 587 nm, 597 nm, 607 nm, 617 nm, and 627 nm. From Giessen et al. (1996a)

To investigate the dephasing in the case of strong confinement, a three beam experiment was performed at small CdSe quantum dots (Giessen et al. 1996a). The scheme of spectral hole burning in the gain spectrum is presented in Fig. 5.22. First, a strong pump beam 1 at 560 nm creates the gain. After 5 ps, when the gain is completely built up, a second pump beam 2, tunable through the gain spectrum and with 160 fs pulse width, is sent to the sample. A broad-band probe beam tests the spectral changes in the gain spectrum at different delay times, ranging from 0 fs to 10 ps. From that type of experiment, one gets direct evidence (i) whether an inversion actually has been produced, (ii) that the mechanism of the gain can only be explained by a coupled level system, and (iii) how fast the dephasing in the gain region is. In particular, we are interested in the information (iii) which can be obtained by analyzing the width of the burnt hole.

Figure 5.23 shows the differential absorption signal of the gain spectrum $[-\Delta\alpha d = (\alpha d$ with pump 2$) - (\alpha d$ without pump 2$)]$ for different delay times and detuning of the pump pulse 2. Focusing first on series (ii) of Fig. 5.23, the temporal development of the $-\Delta\alpha d$ signal can be followed. At early times, a spectral hole in the gain can be observed around the wavelength of pump 2. The appearance of a reduced gain signal (=induced absorption) clearly proves the existence of gain. Without gain, only bleaching (=reduced absorption) would appear in the spectra. Simultaneously with the spectral hole around pump 2 a second hole forms around 565 nm, which grows even faster as the pump pulse continues to take carriers out of the gain region. This dynamics can be well understood in the model already discussed in Sect. 3.2 to attribute the gain to stimulated transitions out of a system of one- and two-pair states. The strong pump 1 beam creates a gain by populating two-pair states. The photon of the pump 2 causes a stimulated transition under emission of a photon. The now incoming probe photon, instead of seeing gain, can be reabsorbed and again a two-pair state will be populated causing an induced absorption feature around 565 nm. Simultaneously, in the same part of the spectrum (around 565 nm), stimulated transitions from the two-pair states to the one-pair states contribute to the signal. The simultaneous occurrence of the reduced gain and the induced absorption feature supports again the coupled nature of the one-pair–two-pair system. Likewise, it is a strong argument against the involving of trap states in the gain spectrum, since their gain depletion cannot produce spectral holes simultaneously both at the trap energy and at higher energies (for details see e.g. Hu et al. 1996, Giessen et al. 1995, 1996a). The series (i) to (v) show the behavior when tuning pump 2 through the gain region. The spectral hole around pump 2 clearly follows the pump beam whereas the differential gain signal around 565 nm does not shift with varying pump.

In the next step, by a convolution procedure described in Meissner et al. (1993, 1994), the width of the burnt hole around the pump has been calculated for different detuning values and thus the energy dependence of the dephasing time has been determined both for large, bulk-like CdSe-microcrystals and for CdSe quantum dots in the strong confinement range. Although the gain spectra look very similar for the various samples (more than 100 meV broad and stretching from below the absorption edge up to the transparency point), the underlying gain processes are very different. In the large quantum dots the gain formation proceeds in analogy to a bulk semiconductor with bandfilling and the development of an electron–hole plasma. In the strongly confined quantum dots, the gain below the absorption is not due to band gap renormalization but due to biexciton to excited exciton transitions, which have a lower energy than the lowest exciton transition in these quantum dots (see Sect. 3.2).

In Fig. 5.24 the results are compared for bulk-like $CdS_{0.7}Se_{0.3}$ quantum dots and CdSe quantum dots with $R \approx 0.5a_B$[3]. The bulk-like $CdS_{0.3}Se_{0.7}$ sample ($R \gg a_B$) shows a T_2 time increasing from 60 fs to 140 fs when the pump 2 pulse is tuned from 1.97 eV towards the transparency point at 2.2 eV (Fig. 5.24a). This is caused by the strong carrier–carrier interactions for electron–hole pairs feeling a plasma-like environment in the large dots,

---

[3] The reason for using a mixed crystal sample for the bulk material was determined by sample availability and the spectral range provided by the laser equipment. The influence of alloy disorder on the dephasing time consists only in a shortening of the residual background dephasing and does not change the energy dependence towards the transparency point.

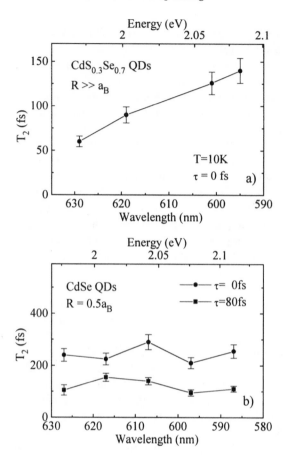

Fig. 5.24. Dephasing times in the gain region of (a) bulk-like $CdS_{0.3}Se_{0.7}$ nanocrystals at a pump-2-probe delay of 0 fs and (b) CdSe quantum dots at pump-2-probe delays of 0 fs (circles) and 80 fs (squares). The lines are a guide to the eye. From Giessen et al. (1996a)

where the electronic states are so close together that they form a band-like structure. Hence a strong excitation above the lowest transition can create a number of excitons in the microcrystallite which interact strongly with each other. However, closer to the transparency point, which corresponds to the top of the Fermi sea in the plasma picture, there are not enough final k-states available for carriers to scatter into. Consequently the scattering rates decrease and the dephasing rates increase substantially, giving rise to longer T_2 times for increasing photon energy in the bulk-like microcrystallites. This result agrees very well with the data obtained by spectral hole burning in the gain region of bulk CdSe (Meissner et al. 1993, 1994).

In zero-dimensional semiconductors like quantum dots, the density of states undergoes a continuous decrease with decreasing radius. Therefore, one would expect a distinctly different behavior of the scattering in systems of reduced dimensionality. A pump slightly above the absorption edge will populate by at most two electrons in the $1s_e$ level. This system, consisting of a zero-electron–hole-pair ground state, a few one-electron–hole-pair (exciton) states, and the corresponding two-electron–hole-pair (biexciton) states, is substantially different from the Fermi-sea concept. All levels in this multi-level system interact in a very similar way with the 'scattering background' and do not give rise to reduced k-state availability close to the transparency point. Therefore, no energy dependence is expected for the dephasing times within the gain region. Indeed, this behavior has been found for the strongly confined CdSe quantum dots with $R = 0.5a_B$. The dephasing time lies around an average T_2 time of 250 fs when tuning the pump between 1.97 eV and 2.12 eV (Fig. 5.24b) and shows a rather constant value across the gain region, in contrast to the continuous increase towards the transparency point in the bulk-like $CdS_{0.3}Se_{0.7}$ sample.

For later delay times we observe a shortening of the dephasing times by a factor of about two in strongly confined CdSe quantum dots (95 fs and 155 fs for the pump-2-probe delay of 0 fs and 80 fs, respectively). Similar behavior has been observed by spectral hole burning in the absorption region of the same strongly confined quantum dots (Woggon et al. 1995d) and explained by a multicomponent dephasing time. On very short time scales (first 100 fs) the different dephasing channels act with different rise times due to finite coupling times for the different processes, such as a delayed onset of dephasing due to lattice vibrations.

6. Trap Processes

In the preceding chapter we have discussed how phonons, interface-related defects, and incoherent electrons or holes can influence the dephasing time T_2. The scope of this chapter is the investigation of the recombination process of the excited electron–hole pairs expressed in terms of the lifetime T_1. In general, a trap can be considered a state which decreases the electron–hole overlap of the excited pair and therefore increases the recombination time. In quantum dots the most likely trap process is the capture of an electron or hole by a local potential in the interface region. However, in quantum dots the interface-related trap processes modify the electronic properties much more than an impurity in the bulk semiconductor usually does. Trapping changes not only the recombination process, but can also cause lattice distortion, alterate the potential barrier, or activate chemical reactions. After the trap process of the optically excited electron or hole, the quantum dot can be considered a completely changed new object.

A recent goal of quantum dot research is high luminescence efficiency. The reduction of the dimensions in real space means that there is a large extension of the electronic states in k space. In bulk materials, transitions at $k = 0$ and $k \neq 0$ (direct and indirect recombination) are clearly distinguishable by different recombination probabilities and different time behavior. In quantum dots, however, the relaxation of the k-conservation rule might have important implications both for radiative transitions and for trapping. For example, in the kinetic models proposed in the literature, one usually neglects all k-conservation rules.

A very crucial parameter for describing the process of localization and trapping is the relation between the spatial extension of the electron and hole wavefunctions and the reach and depth of the local potential fluctuations. The transition from extended to localized electronic states shows different characteristics depending on whether the wave function of the pair states averages over local potential differences or not.

In the following, examples will be given for main mechanisms of trap processes both for the case $R \gg a_\mathrm{B}$ and $R \ll a_\mathrm{B}$. An experimental proof of the influence of interface related traps will be given by comparing the recombination process and the nonlinear optical properties of quantum dots

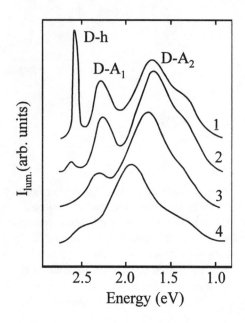

Fig. 6.1. Luminescence spectra of CdS quantum dots of different radii: (1) 28 nm, (2) 7.4 nm, (3) 3.3 nm, (4) 1.5 nm. From Ekimov et al. (1990a)

with different interface configurations. The different kinetic models involving trap processes in quantum dots will be the subject of a critical discussion.

6.1 Localization, Trapping and Transfer

The electronic states responsible for carrier capture should be present in the optical spectra. Looking at the linear absorption, the existence of these bound states can only be guessed in the tail of the absorption towards lower energies. More obvious is the presence of traps in the spectra of luminescence and nonlinear absorption. The way to identify trap states starts from the analysis of the spectra of emission followed by time-, temperature- and intensity-dependent investigation of their dynamics. In Figs. 6.1–6.4 representative experiments have been summarized, illustrating the presence of traps in luminescence (Figs. 6.1 and 6.2), in the luminescence decay (Fig. 6.3) and in the time behavior of the absorption bleaching (Fig. 6.4) for II–VI quantum dots.

Figure 6.1 shows the results of a systematic study of the luminescence spectra for CdS quantum dots carried out by Ekimov et al. (1990a). The sizes of dots used in these measurements cover a wide range. In bulk semiconductors the steady-state luminescence at low temperatures is a useful tool for identifying the nature and concentration of impurities. For quantum dots, only a few systematic luminescence studies exist which include the whole range of sizes starting from bulk up to the strong confinement limit.

Some of the reasons for this are difficulties in the precipitation of dots over such a large range of radii during one and the same growth procedure.

In Fig. 6.1, the luminescence peaks found for the largest sizes of CdS dots ($R = 28$ nm) can be assigned to a donator–hole transition (D–H), and to donator–acceptor-pair transitions (D–A_1) and (D–A_2) in analogy to the bulk material. This classification can be followed up to sizes somewhat above the Bohr radius. In dots of smaller sizes the luminescence no longer resembles the original bulk feature. The line shape is still similar to donator–acceptor-pair transitions, however the nature of these impurities has changed obviously. The luminescence spectrum of the smallest CdS quantum dots with $R = 1.5$ nm is more similar to the luminescence shown later in Fig. 6.3 for small CdSe quantum dots of $R = 2$ nm. A common feature of these steady-state luminescence spectra in small-sized quantum dots is the spectrally broad, strongly red-shifted (~ 700 meV) band. In most of the samples grown without special optimization, the direct 'band-edge' recombination of the electron–hole pair is only visible by a weak, low-efficient luminescence peak, often too small for reasonable detection.

The red-shifted emission may be of different origins such as site-substituted impurities, lattice distortions or surface defects. A site-substitution model is favored if the red-shifted emission is not size-dependent, but sensitive to the chemical preparation method. The small dimensions of the nanocrystals and the preparation techniques even provoke the inclusion of impurities, in particular when using glassy matrices. A site-substitution model was introduced by Misawa et al. (1992) to explain the emission spectrum and decay of CdS nanocrystals precipitated in polymeric films. The authors observed a strong modification of the red luminescence depending on the kind of cadmium salts used during the chemical synthesis. Because these salts differ only in their counter ions [e.g. halogens like $CdCl_2$, CdI_2, or molecule groups $Cd(NO_3)_2$ and $Cd(ClO_4)_2$], the trap states created have been attributed to impurities where the sulfur ion is replaced by the counter ion of the cadmium salt. The replacement proceeds very easily if the size of the counter ion nearly equals the sulfur radius. The sulfur ion radius amounts to 0.104 nm and thus halogen ions of similar sizes can take its place. In contrast, the inclusion of molecular groups more than twice as large causes strong lattice deformations. In this situation, the sulfur atom can hardly be substituted. Hence, the choice of the counter ion of the cadmium salt represents a possibility to control the impurity luminescence. The use of large counter ions in the Cd salts may consequently enhance the band-edge emission.

An example for the correlation of the trap luminescence and the sample preparation technique is shown in Fig. 6.2. Figure 6.2a shows the absorption and luminescence spectrum of CdS nanocrystals synthesized by use of cadmium acetate $Cd(OCOCH_3)_2$ (large radius of the molecule group) and Fig. 6.2b shows the same preparation procedure but making use of $CdCl_2$, the

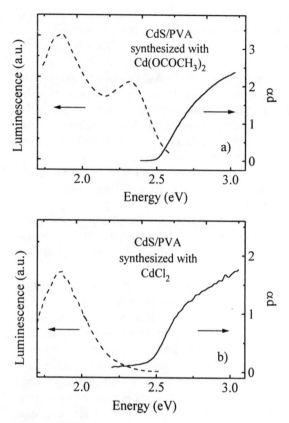

Fig. 6.2. Absorption and luminescence spectra of CdS nanocrystals synthesized with cadmium acetate $Cd(OCOCH_3)_2$ **(a)** and $CdCl_2$ **(b)**. From Sperling et al. (1996)

salt with the smaller chlorine ions. In agreement with the discussion above, the emission spectra exhibit totally different features.

Impurity luminescence owing to site-substitution is usually not influenced by the radius of the nanocrystals and appears also if the grown crystallites reach bulk-like properties. Next we turn to trap mechanisms where the aspect of small sizes comes into play. The influence of the nanocrystal size is predominantly mediated by the varying surface-to-volume ratio.

The Case of Quantum-Dot Sizes $R \ll a_B$. For $R \ll a_B$, it is supposed that the wave function of the electron–hole-pair states averages over all local potential fluctuations. When the radial distributions of the electron or hole wave function have their maxima near the interface, the probability is enhanced for real capture of the carrier in the interface region. Later then, a slow migration of the charged carrier over microscopic distances within the glass matrix might occur. Within this simplified consideration, interface localization, trapping and transfer can be regarded as successive processes and the dot–matrix boundary essentially determines the course of this procedure. In the following we collect arguments and experimental hints to confirm this concept.

Detailed information about the trapping process and the nature of the capture centers can be obtained from time-resolved measurements of the luminescence decay. The results of such experiments are presented in Fig. 6.3 for CdSe (Wind et al. 1994). The luminescence shows a peak near the absorption maximum (2.13 eV) and a broad low-energy shifted band (1.87 eV). The quantum dots were excited at 2.504 eV by use of an excimer laser pumped, quenched transient dye laser system providing 70 ps pulses of maximum fluence of 60 mJ/cm^2. This energy corresponds approximately to resonant excitation into the $1s_e \longrightarrow 1S_{1/2}$ transition. Figures 6.3b and 6.3c give the temporal evolution of the luminescence intensity I versus time on two time scales of 600 ps and 60 ns measured at the two maxima of the luminescence spectrum from Fig. 6.3a. A very fast component has been found at 2.13 eV near the absorption edge. It decays on a picosecond time-scale not resolvable within the exciting pulse-width of 70 ps (Fig. 6.3b). A deconvolution of the luminescence profile with the laser has shown that the maximum of the luminescence is not delayed with respect to the laser maximum. This behavior is an indication that the recombination process has a short component of some tens of picoseconds only. This fast picosecond component close to the absorption edge is usually attributed to the direct recombination of the excited electron–hole pairs. The broad emission band at 1.87 eV dominates the luminescence in the nanosecond time range (Fig. 6.3c). Here the very fast component is missing. This part of the luminescence spectrum is assumed to be related to the recombination arising from traps in the semiconductor–matrix interface region. After the excitation, carriers may rapidly be transferred into these states either directly from the resonantly excited pair state or via the ground state after relaxation. The temporal evolution of the nanosecond component cannot be explained by a simple monoexponential decay. At least two components are involved in the recombination process as obtained from a two-exponential fit giving recombination times $\tau_1 = 5$ ns and $\tau_2 = 120$ ns. The occurrence of more than one decay time is understandable considering the dot–matrix boundary as a source of a great variety of trapping centers.

Finally, Fig. 6.4 shows a new phenomenon observed in II–VI doped glasses called 'photodarkening'. In semiconductor-doped glasses, the photoinduced darkening comprises those phenomena detected after strong laser exposure, such as a decrease of the nonlinear optical response and of the luminescence efficiency, a faster recovery of the absorption bleaching, and the disappearance of the slow luminescence component in the red part of the spectra.

In Fig. 6.4 the time behavior of the absorption bleaching has been investigated as a function of the number of laser shots. At resonant excitation with an energy of 55 μJ/pulse, the maximum transmission change decreases

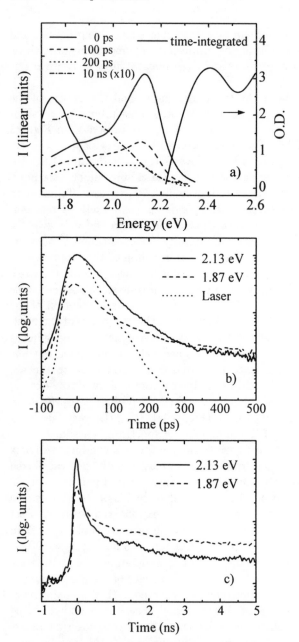

Fig. 6.3. (a) Absorption and time-resolved luminescence spectra measured at different delay times after excitation at 2.504 eV of CdSe quantum dots with $R = 2$ nm; (b) photoluminescence intensity I versus time showing the luminescence decay for two selected photon energies of the luminescence spectrum shown in (a) on a 600 ps time scale (plus the laser profile) and (c) on a 60 ns time scale. From Wind et al. (1994)

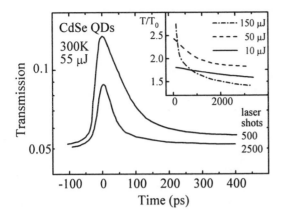

Fig. 6.4. Change of the nonlinear transmission and the decay time of a fresh CdSe quantum dot sample with increasing number of laser shots demonstrating the photodarkening effect

with increasing number of laser shots and the recovery time becomes faster. The inset shows the darkening process with the exciting laser energy as a parameter. The higher the energy dose by which the sample is charged, the stronger the darkening effect.

The photodarkening of $CdS_{1-x}Se_x$-doped glasses was described for the first time by Roussignol et al. (1987a) and attributed to a photochemical process involving impurities probably near the interface. When taking into account the photodarkening, the contradictory results concerning the size and speed of the nonlinearities reported by several groups (see Sect. 6.3) could be explained. The search for the origin of photodarkening and permanent photochemical changes have been carried on, for example, by Kang et al. (1994), Naoe et al. (1994), Masumoto et al. (1995), and Artemyev et al. (1995).

Kang et al. (1994) investigated CdS quantum dot samples with similar particle sizes but prepared by different sol-gel and thermal annealing techniques. The spectral behavior found for the photodarkening showed the same characteristics as known from the calculation of the quantum confined Stark-effect in quantum dots (see Chap. 7). Therefore, to explain the photodarkening, the ejection of photoexcited carriers out of the quantum dots into trap sites at the interface has been suggested, there creating local electric fields which in turn change the internal electronic states of the quantum dot. The photodarkening effect could be reduced if the glass matrix was changed from boro-silicate glasses to silica sol-gels, thus diminishing the concentration of trap sites in the matrix. Persistent spectral hole burning (and its application for data storage) has been studied by Naoe et al. (1994) and Masumoto et al. (1995). The long-living change in nonlinear absorption has been attributed to carrier capture in different ground state configurations of the dot-matrix system, partly produced by the laser illumination itself.

The experimental results of Fig. 6.1–6.4 can be applied to most of the II–VI-doped glasses presented in the literature. Over a long time no microscopic knowledge about the glass–semiconductor interface existed and the

assignment of the experimental findings to traps was only a phenomenological explanation. Different possibilities exist to prove that traps at the
interface participate in the recombination dynamics and the long-time behavior. Some of them include the controlled change of the interface and of the
matrix composition, the evidence of photoproducts within the matrix, or the
development of kinetic models which involve the interface traps and fit well
to the decay behavior.

In glasses, experiments which give a direct comparison of nonlinear optics before and after a controlled interface modification are rare. Only in the
case of quantum dots grown in organic environments is the surface capping
by different organic groups well developed and an optimization of nonlinear
dynamics by interface engineering conceivable (Bawendi et al. 1992; Wang
1991b). Likewise, in ZnCdS quantum dots in aqueous solution, electron ejection out of the semiconductor has been detected by the yield of photoproducts
absorbing at 616 nm, a wavelength well-known and typical for hydrated electrons (Ernsting et al. 1990).

For semiconductor-doped glasses, Ekimov and Efros (1988) proved over-
barrier transitions of electrons out of the quantum dot into the glass by
measuring the thermo-stimulated luminescence. In CdSe-doped glasses, the
intensity- and time-dependent nonlinear absorption was investigated after
controlled surface passivation of small CdSe quantum dots by hydrogenation
(Woggon et al. 1990, 1992). The capture of electrons in long-living traps
outside the dot was also established by MacDonald and Lawandy (1993). It
can explain second harmonic generation in semiconductor-doped glasses. The
creation of a permanent dc-electric field resulting from the charge separation
is exploited to introduce $\chi^{(2)}$-related effects.

The importance of deep trap levels in the interface region has also been
confirmed for dot–matrix systems based on III–V semiconductors. A theoretical treatment of the coupled quantum–dot matrix system was carried out
by Sercel (1995) for $InGa_x As_{1-x}$ quantum dots embedded in GaAs. A deep
trap level in the interface, coupled to the internal electronic dot levels via
multiphonon-assisted tunneling is able to provide a fast channel for energy
relaxation.

In Sect. 6.2 we summarize and give a short overview of the kinetic models
applied in the literature to describe the recombination dynamics by involving
trap states.

The Case of Quantum-Dot Sizes $R \gg a_B$. Supposing next that the
wave function of electrons and holes does not average over local potential
fluctuations at the interface, then a considerable part of the carriers will
relax into localized states which are widely distributed in energy owing to
the different depths of the potential fluctuations across the interface. The
decay dynamics follows a stretched exponential function $\exp\left[-(t/\tau)^\beta\right]$ (see
also Chap. 9) and the microscopic description is a random walk of excited
carriers among the trap sites until a radiative recombination center is found.

Since non-radiative recombination is suppressed, this type of carrier localization may considerably enhance the luminescence efficiency. The stretched
exponential kinetics, high quantum efficiency up to room temperature, and
temperature-dependent activation behavior are common features and typical
signs for carrier localization effects.

Electron–hole-pair localization was considered by Klimov et al. (1996)
and Beadie et al. (1995) to explain the carrier kinetics and temperature-
dependence of luminescence in II–VI-doped glasses with quantum dots of
$R \geq a_B$. The localization effects are of particular importance for quantum
dot structures obtained by epitaxial growth, such as for strain-confined GaAs
dots (Zhang et al. 1995) and InAs dots obtained by self-organized growth on
GaAs substrates (Lubyshev et al. 1996).

When the quantum dot size is increased further, the surface-to-volume
ratio loses its importance. The interior of the dots is relatively undisturbed by
the interface properties and the recombination shows typical features known
from the bulk. In the decay dynamics one regains the intensity-dependent
time constants typical for, e.g., Auger recombination (Ghanassi et al. 1992;
Rougemont et al. 1987; Chepic et al. 1990) and the recombination behavior of
a confined, high-density electron–hole plasma system (Jursenas et al. 1995).

6.2 Kinetic Models

In the following we give a short overview of the kinetic models applied in the
literature to describe the recombination dynamics by involving trap states.
(Examples for the stretched-exponential kinetics will be given in Chap. 9).
Rate equation systems, like those schematically depicted in Fig. 6.5, are generally used for this purpose. For example, in bulk samples the excitation
and generation of electron–hole pairs proceeds from the valence band (1)
to the conduction band (2). Impurities were introduced as near-band traps
for electrons (3) or holes (4). The efficiency of the different channels is characterized by the applied parameters of the cross-sections. Nonradiative and
radiative processes are possible. More complicated considerations involve not
only linear recombination but also quadratic or cubic terms for the rates
of electron–hole pairs, typically for example for Auger recombination. Rate
equations are mostly ambiguous. Therefore, it is not surprising that several
models fit the same experimental results well.

In quantum dots, one likewise starts from three- or four-level systems
for the analysis of the time behavior. The peculiarity in quantum dots
which creates difficulties is the dependence of the optical nonlinearity (or
luminescence) on the laser exposure time, i.e., their decrease in efficiency
and decay time. The models presented for quantum dots can be classified as (i) three- or four-level systems without taking into account photodarkening (Bawendi et al. 1992; Mitsunaga et al. 1988) and (ii) modified three-level systems including photodarkening by either the intensity-

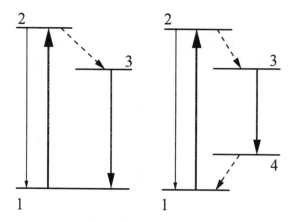

Fig. 6.5. Three- and four-level models

dependent activation of a new trap state (Tomita and Matsuoka 1990; Van Wonterghem et al. 1989) or two-step absorption to excited states followed by the transfer into the glass matrix (Woggon et al. 1990, 1992; Malhotra et al. 1991). In addition, Auger processes followed by trapping have been considered by Ernsting et al. (1990), Ekimov and Efros (1988) and Roussignol et al. (1987b). A three-level system involving only the valence band, the conduction band and a saturable trap state, as seen in Fig. 6.5, fails to explain the photodarkening effect and its long-time behavior.

From their mathematical structure all those models are suited for fitting the peculiarities in the kinetic behavior of quantum dots, which contain terms proportional to the intensity I of the exciting laser, like the number of available trap centers $N(I)$, or by parametrizing the capture efficiency $\sigma(I)$ or the recombination time $\tau(I)$ as function of the intensity I.

Some exemplary ideas will be outlined in more detail below. Intensive and thorough work has been published by Bawendi et al. (1992) on CdSe quantum dots surface-capped by different organic groups. The luminescence spectra obtained for these samples consist of the two characteristic bands of the band-edge related fast and the low-energy shifted slow component with decay times of 100 ps and some μs, respectively. Since photodarkening is not crucial for quantum dots in an organic environment, the long-time behavior does not have to be considered. A three-state model has been proposed with ground state, the direct populated excited states and the trap states populated by radiationless transition from the excited state. To describe the dynamics, a resonance between the internal quantum dot energy states and the surface states has been assumed. The surface band composed of lone pairs on Se atoms is predicted from theory to lie energetically near the valence band top, at least for the size of quantum dots investigated (CdSe with $R = 1.6$ nm). Schanne-Klein et al. (1992) and Hache et al. (1991) expanded these ideas and applied them to glasses, formulating the three-level model with surface and volume states for $CdS_{1-x}Se_x$ quantum dots.

A simple three-level system with valence band, conduction band and trap level was already tested by Mitsunaga et al. (1988) at the beginning of the studies on the quantum dot recombination dynamics in 1988. To explain the photodarkening, these authors proposed the parametrization of the transition rate to the trap level depending on the illumination time. The adequate process could be a chemical change of the traps or a change in their concentration. Using that model with a *decrease* in the trapping rate with time, the acceleration of the decay behavior is well explained by the dominance of direct recombination. However, the small signal efficiency after the photodarkening is disregarded.

The treatment of the photodarkening by parametrizing the trapping efficiency has proved to be a very promising ansatz. This idea has been developed further by Tomita and Matsuoka (1990) and Van Wonterghem et al. (1989). To obtain coincidence with the experimental facts, namely fastening of the recombination and simultaneous decrease of the nonlinearity, an extended three-level system has been successfully applied now involving a fourth transition. The fourth transition is characterized by an increasing capture efficiency over time. Its inclusion opens a new channel of decay which depopulates the excited state more efficiently than the direct recombination to the ground state. In contrast to the first model of Mitsunaga et al. (1988) this transition is *not* saturable, moreover its efficiency grows with time (Tomita and Matsuoka 1990; Van Wonterghem et al. 1989). For the microscopic nature of that new recombination channel a second minimum in the trap potential curve has been proposed by Tomita and Matsuoka (1990), based on the temperature dependence of the photodarkening observed in their samples. Since the photodarkening is slower at low temperatures the process depends on thermal activation, which means that a potential barrier has to be overcome. Contrary to that explanation, Van Wonterghem et al. (1989) did not find any temperature influence or any other threshold behavior of the photodarkening. Therefore they use a second trap created by illumination for the additional new state. The trapping rate into that new center is proportional to its concentration which increases with laser exposure.

The last group of models relate the trap directly to the interface (for example Woggon et al. 1990, 1992; Malhotra et al. 1992). Its population is more probably from the excited states of the quantum dot and trapping can be followed by transfer processes into the glass. By means of a two-step excitation process (photoassisted trapping) electrons and holes populate states within the semiconductor–matrix interface region. The concentration of states can be changed by illumination or is per se very high and therefore not saturable. From its mathematical structure, the description by two-step absorption is very similar to the Auger process within a four-particle system as proposed by Ernsting et al. (1990), Ekimov and Efros (1988), and Roussignol et al. (1987b).

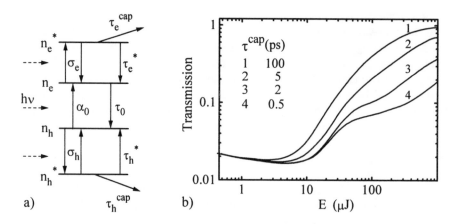

Fig. 6.6. Kinetic models including capture and transfer processes both for electrons and holes (**a**), and calculated transmission versus pump energy (**b**) for resonant excitation and varying the carrier capture times with $\tau_e^{cap} = \tau_h^{cap}$, $\sigma_e = 10\,\sigma_h$, $\tau_0 = 200$ ps, and $\alpha_0 d = 4$. From Woggon et al. (1990, 1992)

To illustrate this last type of models, schematically illustrated in Fig. 6.6, in the following we will discuss an experimental example demonstrating the influence of the interface on the nonlinear optical properties of small CdSe quantum dots. A simple model will be proposed suited to account for trapping and transfer via the interface. The cross-section as the fit parameter is found to be correlated with the chemical nature of the interface configuration. At resonant excitation the transfer of excited carriers into traps of the semiconductor–matrix interface is supposed, with the consequence of smaller bleaching, as expected in the ideal two-level system. To calculate the intensity dependence of the absorption change, the model in Fig. 6.6a has been used, involving discrete quantum dot energy states and resonant excitation of electron–hole pairs. The processes occurring are (i) the absorption bleaching of the transition between the lowest electron and hole levels, n_e and n_h, respectively, (ii) the population-dependent transition of electrons or holes to higher states, n_e^* and n_h^*, (induced absorption with the cross-sections $\sigma_{e,h}$) and (iii) the carrier capture and transfer into interface or glass traps described by the capture time $\tau_{e,h}^{cap}$. Starting from a system of coupled differential equations according to Fig. 6.6a, the transmission has been calculated in relation to the pump intensity with the capture time $\tau_{e,h}^{cap}$ as the parameter, and the results are shown in Fig. 6.6b. For the calculations, the cross-section for the absorption of electrons has been assumed to be different from that of the holes. The results in Fig. 6.6b have been computed with $\sigma_e = 10\,\sigma_h$ and $\tau_0 = 200$ ps; τ_0 is the recombination time of the direct electron–hole pair transition and would correspond to the lifetime T_1 in the undisturbed two-level system. The value of 200 ps has been chosen for the fitting procedure in compliance with values measured in the bulk material.

If the capture time $\tau_{e,h}^{cap}$ is of the same order of magnitude as τ_0 (curve 1 in Fig. 6.6b with $\tau_0 = 200$ ps and $\tau_{e,h}^{cap} = 100$ ps), the intensity-dependent change in transmission is to some extent two-level-like. Transmission $T = 1$ can be reached at reasonable laser excitation intensities. Making the transfer process more efficient by lowering $\tau_{e,h}^{cap}$, a complete saturation can hardly be achieved even at the highest intensities. For capture times below $\tau_{e,h}^{cap} \approx 2$ ps, a new characteristic behavior has been found. Over nearly one order of magnitude of the excitation intensity, the saturation is constant caused by the influence of the induced absorption process. The transient constancy of the bleaching is a consequence of the fact that both kinds of carriers are captured and transferred, and the saturation is determined by the carrier type having the larger σ and the smaller τ^{cap}.

To confirm these calculations the absorption change has been investigated after controlled surface passivation of small CdSe quantum dots by hydrogenation. The CdSe-doped sample was cut into two pieces and one was exposed to a hydrogen atmosphere at 300 °C for 4 hours. The method of hydrogenation of defects is known from the passivation of Si-SiO$_2$ interfaces and a saturation of dangling bonds and of nonbridging oxygen-bonds is expected. To study the absorption saturation and its time dependence, two-stage amplified 17 ps single pulses from a frequency-doubled passive mode-locked YAG laser ($\lambda_{exc} = 539$ nm) were used. The CdSe-quantum dots were excited resonantly and the change in transmission was measured in dependence on intensity for both pieces of the sample. All measurements were carried out after an initial exposure of each sample to 2000 laser shots. In Fig. 6.7 the probe beam transmission versus pump intensity is shown for both samples measured at room temperature and at a fixed delay of the probe beam of 30 ps. For the hydrogenated sample the observed switching factor between the transmission levels at low and high excitation intensity is significantly larger than that of the nonpassivated sample. Obviously, the hydrogenation results in an improvement of the nonlinear optical properties. In Fig. 6.7, additionally two theoretical curves have been plotted, which have been calculated along the model sketched in Fig. 6.6. Fitting the observed results before and after the hydrogen treatment yields qualitative agreement when using $\tau_{e,h}^{cap} = 5$ ps, $\sigma_e = 5 \times 10^{-16}$ cm^2, $\sigma_h = 8 \times 10^{-17}$ cm^2 for the hydrogenated sample, and $\tau_{e,h}^{cap} = 1$ ps, $\sigma_e = 8 \times 10^{-16}$ cm^2, $\sigma_h = 8 \times 10^{-17}$ cm^2 for the nonpassivated one. These values of τ^{cap} and σ reflect an enhancement of trapping in the non-passivated sample. The constancy of the bleaching (Fig. 6.6b) appears in the range of highest intensities used in the experiment.

Another attempt of testing the proposed simple model was made by changing the properties of the glass matrix. The nonlinearity was investigated in CdS quantum dots of similar sizes but embedded in glasses from different melts and annealed under different conditions. In Fig. 6.8 we

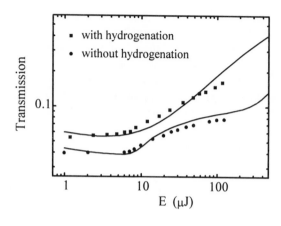

Fig. 6.7. Measured probe beam transmission versus pump energy of the non-passivated (circles) and the surface passivated (squares) CdSe quantum dots samples at $T = 300\,\mathrm{K}$ and the calculated curves following the model of Fig. 6.6. From Woggon et al. (1992)

show the intensity-dependent change in transmission of CdS quantum dots grown at 680 °C over 18 days without a preceding nucleation procedure and, for comparison, the results obtained for the high-quality sample grown at 600 °C/60 h containing the 7.5 nm radius quantum dots.

In both cases the three curves plotted in Fig. 6.8a and 6.8b present the calculated absorption saturation for the ideal two-level system and the fits of the experimental data obtained at 8 K and 300 K according to the model of Fig. 6.6. From the fitting procedure the modified saturation intensity I_S can be derived and used as a parameter suitable for the discussion of differences in the nonlinear optical behavior. A measure of the quality of the sample is the deviation from the ideal behavior expected in the case of two-level saturation. A simple measurement of the intensity-dependent change in transmission can be used to estimate in a proper way the influence of the growth process on the nonlinear optical properties. The observed temperature dependence is interpreted in terms of thermal population of the trap states. Since trap centers are thermally populated at low temperatures, considerable saturation can also be found in the case of low-quality samples.

The model of Fig. 6.6 also provides an explanation for the long-time behavior of photodarkening, which we are going to demonstrate by calculating its time-dependence. For the discussion of the mutual influence of σ and τ^{cap}, the calculation of the transmission change is shown in Fig. 6.9. As we can see, the curves obtained show exactly the typical decay from Fig. 6.4. The photoinduced modifications have been introduced by parametrizing the cross-section σ and the capture time τ^{cap}. Curves 1 and 2 were obtained by assuming capture times $\tau_{\mathrm{e,h}}^{\mathrm{cap}} = 5$ and 1 ps, respectively, and $\sigma_{\mathrm{e,h}} = 8 \cdot 10^{-17}\,\mathrm{cm}^2$. The maximum bleaching value is smaller for $\tau^{\mathrm{cap}} = 1\,\mathrm{ps}$, corresponding to an efficient carrier capture in a photodarkened sample. The decay of the fast components is similar and is caused by the very short τ^{cap}-values chosen. A shortening of the fast component – as observed in Fig. 6.4 – also could only be achieved by

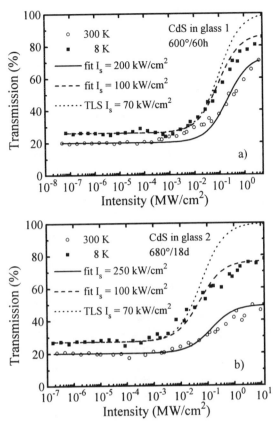

Fig. 6.8. Intensity-dependent change in transmission in a single-beam experiment for CdS quantum dots in different glass matrix materials measured at 300 K and 8 K and compared with the ideal two-level system. From Sperling et al. (1996)

an increase in the cross-section σ (increasing induced absorption), showing that σ must be modified by illumination, too.

In summary, strong evidence for fast trapping into interface traps in the ps- and sub-ps time scale has been found. In the model discussed in more detail, modifications of the dot–matrix interface region are reflected by parametrizing σ and τ^{cap}. This assumption corresponds to changes of interface-related parameters like, for example, the formation or alteration of traps or modifications of the potential barrier. Experiments carried out by modifying the interface properties reflected the predicted behavior. At very short capture times τ^{cap}, the dynamics of the component attributed to the direct recombination can also be influenced by interface-related trapping.

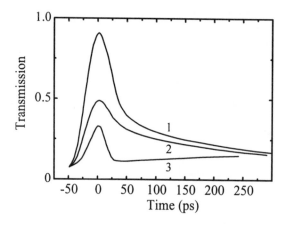

Fig. 6.9. Time behavior of the quantum dot transmission after ps excitation in a two-beam experiment, calculated for capture times $\tau_{e,h}^{cap} = 5\,\mathrm{ps}$ (curve 1), and $1\,\mathrm{ps}$ (curve 2), and a cross-section $\sigma_{e,h} = 8 \times 10^{-17}\,\mathrm{cm^2}$. Curve 3 is calculated with $\tau_{e,h}^{cap} = 1\,\mathrm{ps}$ and $\sigma_{e,h} = 4 \times 10^{-17}\,\mathrm{cm^2}$. From Woggon et al. (1992)

6.3 Trap Processes and Nonlinear Optical Properties

As will be outlined in Chap. 10, nonlinear optical properties have become of increasing technological importance. The characterization and selection of glass compositions and semiconductor dopants demand a consensus on a physical quantity useful for comparison. In the special case of designing devices, there exists a so-called 'figure of merit' to characterize the potential of the optical nonlinearity for application. The concept which generally succeeds is the evaluation of the nonlinear optical properties by means of the nonlinear susceptibility $\chi^{(3)}$ for the different materials. Indeed, the mechanism determining the nonlinear susceptibility is actually substantially different for the various materials.

An exact treatment of the optical nonlinearity in quantum dots yielding the $\chi^{(3)}$ value has to start from the quantum mechanical calculation of the polarization by using eigenstates and eigenvalues. As we have shown in Chap. 3, the energy states and wave function vary strongly with the sizes of the quantum dots, and different approximation schemes have been applied. For the same reason, the theoretical proposals modelling the nonlinearity are very sensitive with respect to the approach used. Treating the light–matter interaction in the frame of classical electrodynamics within the $\chi^{(3)}$ concept, nonlinear optical effects are usually introduced by the power expansion of the polarization based on the anharmonic oscillator model

$$\mathbf{P} = \epsilon_0 \chi \mathbf{E} + \epsilon_0 \chi^{(2)} \mathbf{E}^2 + \epsilon_0 \chi^{(3)} \mathbf{E}^3 + \dots . \tag{6.1}$$

The applicability of the anharmonic oscillator model to semiconductors is limited. The nonlinearities are most often resonant and caused by changes in the (incoherent) carrier population. In spite of this limitation, one uses the third-order susceptibility to compare the nonlinearities in the different materials.

Exploiting the intensity-dependent change in the refraction coefficient, third-order nonlinearities can be determined by many different processes, like the third harmonic generation, Raman and Brillouin scattering and a large number of three- and four-wave mixing processes, the most common of which is the degenerate four wave mixing (Gibbs 1985). To compare the different materials one measures the nonlinear change in the refraction coefficient n_2 defined by

$$n(\omega) = n_0(\omega) + \Delta n(\omega) = n_0(\omega) + n_2\,I \qquad (6.2)$$

and uses the following relation between $\chi^{(3)}$ and n_2

$$n_2\left[\frac{cm^2}{kW}\right] = \frac{1}{3}\left(\frac{4\,\pi}{n_0(\omega)}\right)^2 \chi^{(3)}[e.s.u] \qquad (6.3)$$

to obtain the nonlinear susceptibility. For the proper transformation of SI-units into e.s.u. see Gibbs (1985).

In the following, an example of a standard setup for the determination of the nonlinear refraction coefficient and of the $\chi^{(3)}$ value will be given, the so-called self-diffraction experiment. To obtain an estimate for the refractive index change, the self-diffraction has been measured in a thin-grating approximation ($d \ll 2\Lambda^2/\lambda$, d thickness of the grating, Λ the grating constant and λ the wavelength of the light) and plotted in Fig. 6.10. The two coherent beams, produced by a beam splitter from an excimer laser pumped dye laser, intersect on the sample under a small angle. The transmitted and diffracted orders are detected time-integrated by an optical multichannel analyzer. The efficiency of a thin grating is correlated with the nonlinear change in the refractive index in simplest approximation by

$$\eta = \frac{I_1^+}{I_T} \approx \left|\frac{\pi\,\Delta n\,d}{\lambda}\right|^2, \qquad (6.4)$$

where I_1^+ is the intensity of the first diffracted order, I_T the transmitted intensity of the pump beam (zeroth order intensity), d the thickness of the grating and λ the wavelength of the probing laser. The use of the transmitted instead of the incident intensity in the calculation of the efficiency allows us to take absorption losses approximately into account.

To demonstrate the influence of the host material, the diffraction efficiencies have been compared in Fig. 6.10 for CdS quantum dots of larger sizes but embedded (i) in a glass and (ii) in polyvinyl alcohol (PVA). For the sample with the polymer film as the host, considerably larger diffraction efficiencies can be seen. A value of η of the order of $\sim 10^{-2}$ already permits the observation of the first diffracted order by the eye. To compare within the $\chi^{(3)}$ concept, the nonlinear change in refraction index has been calculated in e.s.u for the two samples according to Eq. (6.2) and (6.3) [see also Gibbs (1985)]. For the sample embedded in PVA a value of $\chi^{(3)} = 3.2 \times 10^{-8}$ e.s.u. has been obtained which is twice the value of the sample with glass being the host material, namely $\chi^{(3)} = 1.6 \times 10^{-8}$ e.s.u.

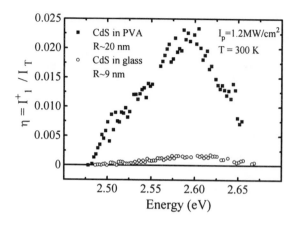

Fig. 6.10. Diffraction efficiency of large-size CdS quantum dots in glass (circles) compared to the embedding in a polymer matrix (squares). From Sperling et al. (1996)

We present in Table 6.1 a summary of the experimentally determined values of $\chi^{(3)}$ (given in e.s.u.) cited in the literature. Keeping in mind that the nonlinearity varies strongly with size and excitation energy (resonant/nonresonant excitation), such a classification is valuable for engineering and technological aspects but is not very instructive with respect to the confinement-induced changes of the optical nonlinearities.

The experimental values reported for $\chi^{(3)}$ in CdS_xSe_{1-x}-doped glasses vary by four orders of magnitude. Saltiel et al. (1990) found that even similar glasses from different or the same manufacturer show a large spread in their value of $\chi^{(3)}$ depending upon the excitation intensity and the exposure history. Although interesting theoretical work about the confinement-induced change of the optical nonlinearities has been published, in the experiments the trapping covers the information about the recombination behavior of the direct transitions. A decreasing optical nonlinearity is often only simulated by the decrease in the recombination time τ changing the figure of merit $F = |\chi^{(3)}|/\alpha\tau$ with α the absorption coefficient.

From the discussion of the phenomena of photodarkening we have seen that the direct recombination process of the quantum confined electron–hole pair is difficult to access. The decision about the usefulness or the harmful character of traps is most likely connected with the understanding of trap chemistry itself.

Table 6.1. Comparison of the nonlinear susceptibility of different $CdS_{1-x}Se_x$ - doped materials; DFWM – degenerate four wave mixing, IF – interferometry, THG – third harmonic generation, PP – pump and probe spectroscopy, res – resonant, nonres – nonresonant

Material/Host		Method/ Excitation	$\chi^{(3)}$ in e.s.u.	Ref.
$CdS_{0.9}Se_{0.1}$/glass	CS 3-68	DFWM res	$1.3 \cdot 10^{-8}$	Jain and Lind (1983)
CdSSe/glass	OG 530	IF nonres	$2 \cdot 10^{-10}$	Danielzik et al. (1985)
$CdS_{0.9}Se_{0.1}$/glass	CS 3-69	IF res	$\sim 10^{-9}$	Olbright and Peyghambarian (1986)
CdSSe/glass	Y-50	DFWM res	$1.3 \cdot 10^{-9}$	Yumoto et al. (1987)
CdSSe/glass	OG 570	DFWM res	$\sim 10^{-8}$	Roussignol et al. (1987a)
CdSSe/glass	RG 630	DFWM res	$\sim 10^{-7}$	Remillard et al. (1989)
CdS/polymer	$R = 1.5$–3 nm	THG res	$0.25 \cdot 10^{-10}$ $- 3.3 \cdot 10^{-10}$	Lap-Tak et al. (1989)
$CdS_{0.9}Se_{0.1}$/glass	CS 3-68	DFWM res	$5 \cdot 10^{-8}$	Saltiel et al. (1990)
$CdS_{0.9}Se_{0.1}$/glass	OG 530	DFWM res	$0.03 \cdot 10^{-8}$ $- 3 \cdot 10^{-8}$	Saltiel et al. (1990)
CdSSe/glass	CS 3-69	DFWM nonres	$6 \cdot 10^{-12}$	Acioli et al. (1990b)
$CdS_{0.12}Se_{0.88}$/glass	$R = 1$–10 nm	DFWM nonres	$10^{-11} -$ 10^{-9}	Shinojima et al. (1992)
CdS/PbS	sandwiched	DFWM res	$1.6 \cdot 10^{-9}$	Spanhel et al. (1992b)
CdSSe/glass	OG 550	DFWM res	$\sim 10^{-9}$	Uhrig et al. (1992)
CdSe/glass	$R = 3$–6 nm	PP res	$6 \cdot 10^{-8}$	Vandyshev et al. (1992)
CdS/PVA	$R \approx 20$ nm	DFWM res	$3.2 \cdot 10^{-8}$	Fig. 6.10 this work

7. Effects of Static External Fields

In the following, we consider the influence of static electric and magnetic fields on the electron–hole pair states of three-dimensionally confined quantum structures. In two-dimensional quantum wells (QW) and superlattices (SL), an external electric field yields the quantum-confined Stark effect connected with a strong red-shift of the excitonic absorption peak (Miller et al. 1984, 1988a). The field ionization of the confined exciton states is prevented by the potential formed by the barrier layers. This field-induced shift of the sharp resonances results in good modulation of the absorption coefficient α, already observable at room temperature. Therefore, it seems reasonable to look for quasi zero-dimensional systems in a next step.

In this chapter we discuss possible mechanisms providing maximum $\Delta\alpha/\alpha$-values for three-dimensionally confined II–VI quantum dots. The theoretical problem of the electric field action has been shown to be very complicated and solvable only by strong simplifications. The theoretical predictions will be summarized shortly. Then we turn to the experimental requirements for their observation. The gain of large electro-optic modulation is essentially a technological problem and a question of optimization.

In the case of magnetic fields, up to the present time only a few observations have been reported, mostly investigating III–V quantum dots. The results obtained are more interesting for basic research and cited here for the sake of completeness.

Although we want to confine ourselves to the treatment of optical properties, we touch the field of transport properties of low-dimensional structures in the last part.

7.1 Electric Field Effects

Most interesting for novel electro-optical devices are semiconducting materials which show large relative changes in the optical density $\Delta\alpha/\alpha$ at comparatively low field strengths ($\sim 10^4$ V/cm) and at room temperature.

Early studies of excitonic effects and of the electronic band peculiarities in bulk II–VI materials, e.g. in CdS, revealed values of $\Delta\alpha/\alpha$ of some percent at room temperature and at fields strengths of 10^4 to 10^5 V/cm in electro-optic experiments (Snavely 1968; Gutsche and Lange 1967; Cardona et al.

1967; Perov et al. 1969; Hase and Onuki 1970). The underlying physical mechanism in bulk CdS to some extent is the excitonic Stark effect, but the broadening and the rise of the absorption tail below the band gap is mainly caused by the field ionization of the discrete exciton states connected with contributions from Franz–Keldysh oscillations (Dow and Redfield 1970; Franz 1958; Keldysh 1958).

Considering II–VI quantum dots, we look in a first step for theoretical concepts predicting a large electro-optic response. To solve the problem, the one-pair Hamiltonian (3.25) has now to be completed by the electric field term

$$\hat{H}_{elstat} = -eFr_{e,h} \,, \tag{7.1}$$

where F is the static electric field strength. An exact treatment should involve both the finite barriers, the Coulomb interaction, and the surface polarization correction, besides the additional electric field term. Since even without the electric field, the Hamiltonian is sufficiently complicated, the electric field problem has been treated so far with substantial simplifications only.

Useful approximations reported in the literature are the consideration of infinite barriers and (i) omitting the Coulomb term at first or (ii) taking into account the Coulomb interaction between the confined electrons and holes. The approach of (i) results in a problem similar to the Franz–Keldysh effect observed in bulk semiconductors for the quasi-free electron or hole in a periodic crystal lattice. In bulk-material the inclusion of the Coulomb potential (ii) corresponds to the exciton and the excitonic Stark effect, which can be treated in analogy to the atomic Stark effect known from textbooks.

Introducing an infinite confining potential barrier in all three dimensions, the classification in the Franz–Keldysh or Stark mechanism is no longer unambiguous. The quantum dot already provides bound states independent of the contribution arising from the Coulomb term. Also, there is no state which would be adequate to a free electron state or hole state like in the bulk. Additionally, considering finite barriers, tunneling through the barriers is very likely and the electrons and holes are no longer confined in their motion, in particular at very high field strengths. The particle motion outside the dot is limited by the low mobility of the carriers in the disordered matrix material.

When discussing the action of the electric field in quantum dots, one is interested to find the conditions which produce field ionization, here related to the bound states of a three-dimensionally confined system (described with or without considering the Coulomb potential). In this context, the problem is sometimes denoted as the quantum-confined Stark effect of quantum dots. However, in a certain range of parameters (finite potential well and high electric field-strength) the electric field-induced spectral changes have to be explained considering both contributions from the quantum-confined Stark and from the Franz–Keldysh effect.

The different theories have to be analyzed with respect to field-induced energy changes of the bound states in the quantum dot in order to use their

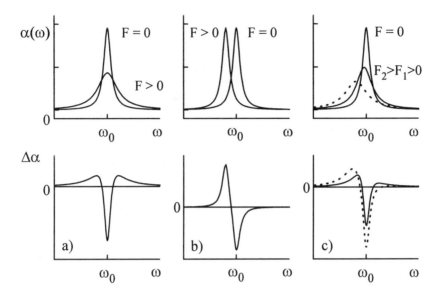

Fig. 7.1. $\Delta\alpha$ spectra (field$_{ON}$ - field$_{OFF}$) expected for dominant line broadening (a), line shift (b), and a combination of both (c) in the case of a single Lorentzian resonance

absorption shift, broadening and loss of oscillator strengths for electro-optic modulation, bearing in mind the aspect of application.

Tendencies for the electric field behavior of II–VI quantum dots can already be derived from the calculations performed for III–V quantum dots. The infinite-barrier problem within a cubic quantum box was investigated by Miller et al. (1988b) by neglecting the Coulomb potential with the main result of a shift of the ground state to lower energies and a redistribution of oscillator strength to formerly forbidden transitions. Assuming infinite barriers, a broadening of the absorption lines is hard to achieve. In contrast, using finite potential barriers, the gradual broadening of the shifting resonances can be pursued up to the ionization threshold, for example, in calculations made by Chiba and Onishi (1988). There the Coulomb potential was applied and the problem solved within the approximation of a fixed center-of-mass at the center of the sphere.

When searching for experiments to examine the theory, one is confronted by the problem of line-shape analysis. In the case of well-separated eigenstates, i.e. single resonances in the optical spectra, Stark spectroscopy can be used to determine line-shift and broadening as illustrated in Fig. 7.1. From the $\Delta\alpha$-spectrum obtained after subtracting the absorption signals with and without electric field, a so-called 'three-line' spectrum can be derived if the increase of line-width dominates, or a 'two-line' spectrum if the center frequency shifts towards smaller values. The line-shape analysis is no longer unambiguous if several states are very close in their energies, if energy separation and line-widths are of same order of magnitude, and if both line-shifts and line-broadening appear. Bearing this aspect in mind, complex behavior of the electric-field induced absorption change is expected for quantum dots.

A detailed study of Stark spectroscopy for that case is given by Sacra et al. (1995) choosing CdSe quantum dots as an example. It has been shown that the broad line-widths of the individual states and the overlapping of transitions coupled by the electric fields give rise to substantial deviations from the known line-shape analysis. A mutual cancellation of shift and broadening effects falsifies the line shape of the single-state $\Delta\alpha$-spectrum considerably. If inhomogeneous broadening is also involved in the zero-field absorption, the $\Delta\alpha$ spectrum may very rapidly lose its unambiguity.

The electric field effects of quantum dots made from II–VI semiconductors have been investigated both experimentally and theoretically since 1989, mainly in the framework of the infinite barrier problem. Hache et al. (1989) applied the independent particle model without Coulomb potential [problem (i)] and the F^2-dependence of the line-shift has been described by second order perturbation theory. The experimentally observed change in the lineshape has been attributed to the electric field-induced change of the selection rules for optical transitions. A similar treatment has been proposed by Rossmann et al. (1990) in the case of small dot sizes below the Bohr radius. For large CdS dots the spectral shape of the field-induced absorption changes has been explained by a field-induced Stark shift to lower energies (Henneberger 1992). The field ionization is prevented up to high field strengths, typically several times the ionization field for bulk excitons.

The Coulomb potential [problem (ii)] was taken into account by Ekimov et al. (1990b) when calculating the problem within the framework of the intermediate confinement model (and infinite barriers). The absorption shift to lower energies $\sim F^2$ up to field strengths of about 10^5 V/cm is explained as the result of the quantum-confined Stark effect.

In a more complete manner, Nomura and Kobayashi (1992) involved both the Coulomb and the surface polarization energy and compared the results with and without Coulomb interaction by use of variational calculations. It has been proposed that the fit of the observed line-shapes by a shift and a broadening better explains the experimental results than taking into account the contribution from former forbidden states. The inclusion of the Coulomb interaction in the calculations gives significant corrections for quantum dot sizes as low as $R \leq 0.33\,a_B$ and reduces the expected low-energy shift of the absorption band.

Esch et al. (1990) compared the results with and without Coulomb interaction by investigating CdTe quantum dots of size $R = 0.3a_B$ and using the matrix diagonalization method. The small differences in the results obtained for the two calculations with and without the Coulomb potential justify discussing their experimental results in the limitation of the quantum-confined

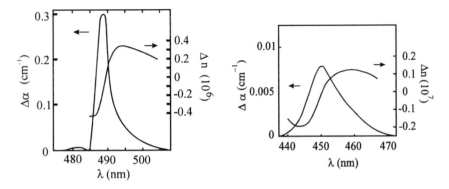

Fig. 7.2. Measured change in the absorption spectra when an electric field strength of 87 kV/cm is applied to CdS nanocrystals of $R = 25$ nm (left) and $R = 1.9$ nm (right), at $T = 77$ K. In addition the calculated change in the refraction index is shown. From Ekimov et al. (1990b)

Franz–Keldysh effect. From these experiments the general conclusion can be drawn that the Coulomb interaction can be ignored in semiconductor quantum dots fulfilling the condition of $R \ll a_B$ (for example $a_B = 7.3$ nm in CdTe).

Experiments in which the sizes and the field strengths vary over a great parameter range were carried out by Cotter et al. (1991). They allowed the observation of the transition from the quantum-confined Stark effect to the Franz–Keldysh effect within a rather small range of quantum dot sizes.

Summarizing the experimental results obtained by applying external electric fields on quantum dots, we can state that the absorption spectra show a small red shift $\sim F^2$ accompanied to some extent by line-broadening. The differences in the data appear in the field strengths necessary for the observation of the effects and in different power laws for the electric field dependence of the absorption change. The experimental studies generally show an increasing $\Delta\alpha$ with increasing dot size at comparable field strengths. The field-induced broadening dominates the absorption change for larger dot sizes.

However, the common results of most investigations are only small changes in the absolute value of $\Delta\alpha/\alpha$ far below 1%, which are often visible only in modulation spectroscopy and at field strengths beyond 10^5 V/cm. These values are distinctly smaller than the data reported for bulk CdS or CdSe at comparable field strengths.

Some examples of the largest absorption changes achieved in the case of embedding the II–VI quantum dots in glass are shown in Figs. 7.2–7.4. Figure 7.2 shows the change measured in the absorption spectra by applying an external electric field of 87 kV/cm to CdS nanocrystals of $R = 25$ nm ($R \gg a_B$) and $R = 1.9$ nm ($R \ll a_B$). In addition, the calculated change in the refraction index is shown. The absence of oscillating structures in the $\Delta\alpha$ spectra which have positive and negative parts is explained by the size

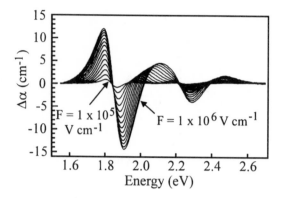

Fig. 7.3. Electroabsorption spectra of $CdS_{0.3}Se_{0.7}$ quantum dots of $R = 8\,nm$, measured at different field strengths and $T = 300\,K$. From Cotter et al. (1991)

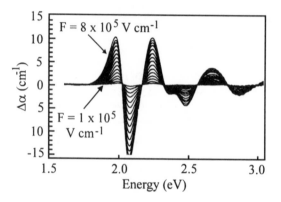

Fig. 7.4. Electroabsorption spectra of $CdS_{0.3}Se_{0.7}$ quantum dots of $R = 2.65\,nm$, measured at different field strengths and $T = 300\,K$. From Cotter et al. (1991)

dispersion and the strong absorption onset at the high-energy absorption edge. Working at low temperatures of $T = 77\,K$, at least 3 % absorption change has been reached in the larger quantum dots of $R = 25\,nm$. Owing to the importance of excitonic effects, the largest changes were expected for quantum dots in the weak confinement case. According to this theory, the experiments show a continuous increase of $\Delta\alpha$ with size.

Figures 7.3 and 7.4 have been taken from Cotter et al. (1991) to demonstrate two typical examples of the different spectral shapes found in the $\Delta\alpha$-spectra. For $CdS_{0.3}Se_{0.7}$ nanocrystals of large radii ($R = 8\,nm$), the electroabsorptive change at different values of F up to $F = 10^6\,V/cm$ is shown in Fig. 7.3. The field-dependent period of oscillation of $\Delta\alpha$ is typical for the Franz–Keldysh effect known from the bulk material at high electric fields. In contrast, for smaller $CdS_{0.3}Se_{0.7}$ quantum dots of radii $R = 2.65\,nm$ (Fig. 7.4) the spectral shape is less dependent on the electric field strength and more typical for a line-broadening mechanism as expected for the beginning of ionization of excitonic resonances. The work of Cotter is the only one known where the applied electric field goes beyond $10^5\,V/cm$. The measured absorption modulation at these high field strengths is about 10 % at $T = 300\,K$. High extrinsic voltages of several tens of kV are necessary to ensure a high field inside the samples of thicknesses up to some millimeters. The decrease in sample thickness down to the μm range is limited due to the low absorption coefficient caused by the low filling factor of semiconductor quantum dots in the glasses. From the small changes in α it may be supposed that additional depolarization occurs in the experiment, which reduces the electric field strength inside the dots. For instance, only little is known about the interface layer around the quantum dot and the local field strength there, or about the influence of deviations from a spherical shape.

Two problems have turned out to be essential with respect to the application of electro-optic properties of quantum dots embedded in glass: (i) the flatness of the absorption edge in the linear spectra owing to the low filling factor ($p \leq 10^{-3}$) as well as the strong inhomogeneous broadening by the size distribution and (ii) geometrical and technological problems in the realization of electric contacts. A further problem is the at present total neglect of interface polarization in the discussion of the electric field problem, both in theory and experiment. Yet, when investigating strained-layer superlattices, a large electro-optic response has been found caused by the modulation of piezoelectrically generated polarization fields by external electric fields or high densities of laser excited carriers (Smith and Mailhiot 1990; Halsall et al. 1992; Langbein et al. 1995). Therefore, it should be inevitable also to take into account interface-induced internal fields in zero-dimensional systems, in analogy to those produced by the lattice mismatch at the interface in layered, two-dimensional systems.

For the above reasons it seems interesting to investigate microcrystals in organic matrices instead of glasses. The promising nonlinear optical properties of CdS in a polymer film environment have already been demonstrated by Misawa et al. (1991), Brus (1991), Wang et al. (1989), Sekikawa et al. (1992), Colvin and Alivisatos (1992a), and Henglein (1989). The polymer film acts as a stabilizer resulting in a high uniformity in the size of the nanocrystals. To our knowledge some first results have been published about the action of an external electric field on CdS and CdSe embedded in an acrylonitril-styrene copolymer (AS) and in polymethyl-methacrylate (PMMA) (Sekikawa et al. 1992; Colvin and Alivisatos 1992a). Electromodulation spectroscopy of absorption and luminescence has been carried out at CdS microcrystals of radii between 1.5 nm and 3 nm embedded in AS and well-resolved structures are attributed to an excitonic Stark shift in this material (Sekikawa et al. 1992). In small CdSe microcrystals of $R = 2$ nm in PMMA, an electric field-induced broadening is explained by a change in the dipole moment of the excited state with increasing field strength averaged over a random ensemble of dipoles rigidly held in the polymer (Colvin and Alivisatos 1992a).

For the experiments presented in Fig. 7.5, mechanically stable PVA-films have been prepared with CdS of volume fractions between 1 to 3 % and thicknesses of some tens of microns (for details see Woggon et al. 1993a). The

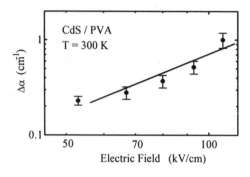

Fig. 7.5. Dependence of the field-induced absorption change on the strength of the electric field, measured at room temperature. From Woggon et al. (1993a)

size of the CdS nanocrystals is in the weak confinement range of $R/a_B \geq 5$, where R is the radius of the crystal and a_B is the excitonic Bohr radius (in CdS $a_B = 2.7$ nm). For the CdS nanocrystals embedded in the polyvinyl alcohol matrix, large values of $\Delta\alpha/\alpha$ up to 7 % have been achieved at room temperature at moderate field strengths of $\sim 5 \times 10^4$ V/cm using planar electrodes and observation in direction of the electric field.

Figure 7.5 shows the dependence of the electric-field-induced absorption change on the applied external field strength. For comparison a quadratic function is plotted as a straight line in the log–log plot (Woggon et al. 1993a). This result shown in Fig. 7.5 allows us to discuss the experiments in terms of the quantum-confined Stark effect as already proposed by Ekimov et al. (1990b), Henneberger et al. (1992), and Nomura and Kobayashi (1992a) for II–VI quantum dots in glasses. The excitonic effects are not destroyed by the electrical field even for field strengths larger than the ionizing field strength of the bulk semiconductor of $\approx 10^4$ V/cm in CdS (Dow and Redfield 1970). The field ionization is suppressed because the exciton is confined by the surrounding potential barrier. The larger change in α can easily be explained by the higher filling factor and by an optimum contact configuration resulting in a homogeneous field and an efficient potential drop over the microcrystal.

In conclusion, to optimize the electric field-induced absorption modulation we have to look for (i) well-defined sizes and thus a good knowledge of the confinement regime, for example, for the estimation of the influence of Coulomb interaction between electron and hole on the electro-optic properties, (ii) a sharp absorption edge or a narrow absorption peak for optimum contrast of modulation, (iii) a volume fraction of the semiconductor quantum dots in the range of 1–10 % to decrease the layer thickness down to some tens of microns, thus ensuring a high electric field at moderate voltages and a reasonable absorption coefficient, and (iv) no (or low) electrical conductivity, realized by an insulating matrix and by keeping an upper limit for the volume fraction of semiconductor dots to prevent a mutual contact or percolation.

Electroluminescence produced by external electric voltage is a further field of intensive research, in particular under the aspect of application possibilities. For the discussion of that problem we refer to Chap. 10.

7.2 Magnetic Field Effects

As discussed in Sect. 7.1 and Chap. 3, electro-optic and nonlinear optical measurements are an efficient tool for determining the quantum dot energy levels in the case of strongly broadened linear absorption more precisely. Recently, Japanese groups have proposed magneto-optic experiments to analyze in a more detailed manner the energy structure of quantum dots (Ando et al. 1993; Nomura et al. 1993a,b). Besides the use of an additional external field for modulating the optical properties and thus for increasing the sensitivity, the field also reduces the number of degrees of freedoms and hence lifts possible degeneracies.

Let us recall the magnetic field action in bulk semiconductor materials. In the presence of uniform magnetic field B along the z axis the eigenenergies of an electron are determined by

$$E(n, \sigma_z) = (n + \frac{1}{2}) \hbar \omega_c + \frac{\hbar^2 k_z^2}{2m_e} + g\, \sigma_z\, \mu_B\, B \,, \qquad (7.2)$$

with ω_c the cyclotron frequency

$$\omega_c = \frac{eB}{m_e} \qquad (7.3)$$

and μ_B the Bohr magneton

$$\mu_B = \frac{e\hbar}{2m_0} \,. \qquad (7.4)$$

m_e is the mass and e the elementary charge of the electron, g the effective gyromagnetic Landé-factor, and σ_z the z component of the spin operator σ (for more details see for example Ashcroft and Mermin 1976; Madelung 1972). The magnetic field acts on the angular momentum of the electron. The first term of (7.2) corresponds to the interaction with the orbital part and leads to the Landau-quantization as the result of the action of the external magnetic field. The last term is the Zeeman splitting resulting from the spin quantization with its projections $\pm 1/2$.

Let us now consider a quantum-confined system. Whether in quantum dots the spin or the orbital part of (7.2) dominates, depends on the material parameters and on the strength of the magnetic field in relation to the confining potential. The effect of the magnetic field on the quantum dot energy states can be classified in different limits comparing the strength of the magnetic 'confinement' given by λ and the quantum confinement owing to the surrounding potential barrier determined by the radius R. The so-called magnetic length λ

$$\lambda = \sqrt{\frac{\hbar}{eB}} \qquad\qquad (7.5)$$

is a measure for the radius of the classical circular motion of the electron in the magnetic field. For example, to observe the orbit quantization and the Landau levels, the magnetic field has to produce orbits smaller than the cross-sectional dimension of the sample. In k space this means that an electron in a magnetic field cannot be localized better than a region of $\Delta k \approx \sqrt{(eB/\hbar)} = \lambda^{-1}$.

If the magnetic field is weak, i.e $R \ll \lambda$, it only acts on the spin part of the wave function and causes the Zeeman splitting. At very strong magnetic fields with $R \gg \lambda$ one also expects deformations in the orbital angular momentum by the external field, and the Landau quantization appears in the optical spectra. In 2D structures the Landau levels have been found together with the famous quantum Hall effect (for a review see Bastard 1988 and references therein).

In quantum dots the magnetic fields usually available can be considered weak to intermediate for dot sizes around the excitonic Bohr radius. Therefore one starts with the discussion of the spin part and introduces the magnetic field in the quantum dot Hamiltonian (3.25) (neglecting Coulomb interaction) by

$$\hat{H}_{magn} = \frac{e\hbar}{2m_0} (g_1\,\sigma_1 + g_2\,\sigma_2)\,B \qquad\qquad (7.6)$$

where the indices 1 and 2 refer to electron and hole.

An experimental method to investigate the magnetic field effects is the magnetic-circular-dichroism measurement proposed for quantum dots by Ando et al. (1993) and Nomura et al. (1993a,b). It is illustrated in the following based on their original work.

In Fig. 7.6 the principle of magnetic circular dichroism (MCD) and its application to CdS_xSe_{1-x} crystals is shown. The band diagram of Wurtzite type CdS_xSe_{1-x} is schematically drawn in Fig. 7.6a with the A, B, and C bands at zero magnetic field. The magnetic field splits the heavy-hole, light-hole and the spin–orbit split-off band due to the Zeeman effect. If the magnetic field has been applied along the light propagation direction, the allowed transitions are two circularly polarized σ_+ and σ_- transitions different in their energies (Fig. 7.6b,c). The MCD signal is defined as $k_-(E) - k_+(E)$ with k the absorption coefficient. It has its zero-position at the energy E_0 (Fig. 7.6d). When this method is applied to quantum dots, the energy levels can be identified taking advantage of the action of the external magnetic field. However, in contrast to the ideal model of Fig. 7.6d, the line-shape will not be symmetrical because of the different g values, oscillator strengths and overlap of the single confined energy levels.

This method has been applied to commercial filter glasses containing $CdS_{0.28}Se_{0.72}$ quantum dots of average radii of $R = 3.8 \pm 0.75\,nm$ (Ando et al. 1993). The MCD spectrum obtained is shown in Fig. 7.7. The dashed

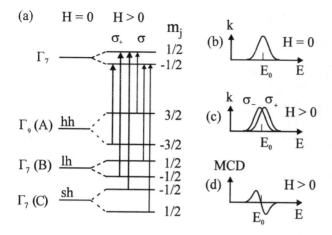

Fig. 7.6. Principle of the magnetic circular dichroism (MCD) measurement in the case of Zeeman splitting in CdS_xSe_{1-x}. The magnetic field is here denoted by H, and m_j corresponds to the spin projection quantum number. σ_+ and σ_- are the two circular polarizations appearing under the external magnetic field (see text). From Ando et al. (1993)

line has been drawn to eliminate the broad positive background. Two peaks at 2.153 eV and 2.247 eV have been resolved and assigned to the hh and lh valence-band states of these $R = 3.8$ nm quantum dots. To compare with the theory, a numerical matrix diagonalization scheme has been carried out using infinite potential barriers, parabolic bands and involving surface polarization effects. Satisfying agreement was achieved.

The Zeeman splittings of the Z_3 exciton and the Z_{12} exciton have been investigated by Nomura et al. (1993a) for CuCl quantum dots of radii $R = 4.6 \pm 1.2$ nm (Fig. 7.8). In CuCl the Z_3 exciton is the energetically lowest exciton state and composed of a conduction-band electron of Γ_6 symmetry and a hole from the Γ_7 valence band. It is therefore twofold degenerate with respect to the spin projection. The linear absorption shows well-resolved the energy positions of the exciton peaks blue shifted by 6 meV (Z_3) and 11 meV (Z_{12}) with respect to the bulk. The Zeeman shifts shown in Fig. 7.8 have been analyzed to determine the g values from the maximum position and the line shape. The g values derived from the experiment are $g = 0.447$ and $g = -0.18$ for the Z_3 and Z_{12} excitons, respectively. The corresponding bulk values were found by Koda et al. (1970) and Suga et al. (1971) to be 0.3 and -0.2. The small differences in the g values have been attributed to the mixing of states with higher l quantum numbers of the orbital angular momentum due to the Coulomb interaction and the nonspherical shape. In addition, the authors propose a superposition of two transitions from the lowest confined

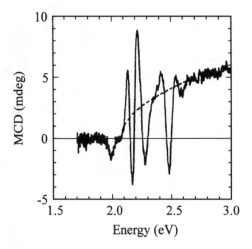

Fig. 7.7. MCD spectrum of a Corning 2-61 filter glass at 15 K ($Cd_{0.28}Se_{0.72}$, $R = 3.8\pm 0.75$ nm) and a magnetic field of 1 T. From Ando et al. (1993)

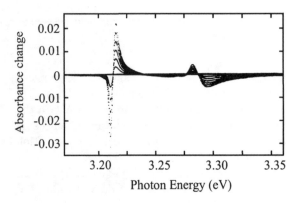

Fig. 7.8. MCD spectrum of CuCl quantum dots of $R = 4.6\pm 1.2$ nm, measured at 10 K and magnetic fields of 1,2,... 6 T. From Nomura et al. (1993a)

states of the two Z_3 and Z_{12} exciton series to explain the line-shape of the second absorption peak.

Gradually increasing the strength of the magnetic field up to 6 Tesla, the MCD-spectra have been measured at $CdS_{0.12}Se_{0.88}$ quantum dots of radius of $R = 3.3$ nm by Nomura et al. (1993b). In this case the strength of the magnetic field is sufficiently high and the approximation of $\lambda \gg R$ is questionable. The magnetic confining length λ and the quantum confinement owing to the radius R reach similar orders of magnitude and the treatment of the magnetic field problem has to involve more sophisticated theories.

As detailed by the authors, the intermediate magnetic field case has been treated by a numerical matrix method including the true valence-band structure and the electron–hole exchange interaction. Because both the spin part and the orbital angular part contribute to the MCD spectra, the energy posi-

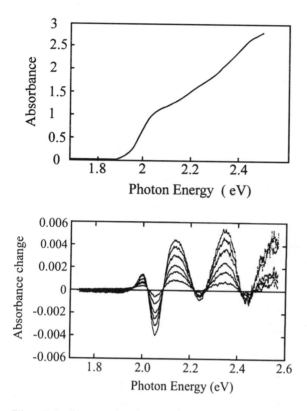

Fig. 7.9. Linear absorption (upper panel) and MCD spectra (lower panel) of CdS$_{0.12}$Se$_{0.88}$ quantum dots, measured at 10 K and magnetic field of 1,2,... 6 T, respectively. From Nomura et al. (1993b)

tion cannot be determined from the zeros of the MCD signal as known from the pure Zeeman-splitting. Also the spectra from the σ_+ and σ_- polarizations are no longer identical owing to the change in the orbital part of the wave functions.

In Fig. 7.9 the linear absorption and the MCD spectra of CdS$_{0.12}$Se$_{0.88}$ quantum dots, measured at 10 K and magnetic fields of 1, 2 ... 6 T are shown (Nomura et al. 1993b). From electroabsorption measurements from the same sample (Nomura and Kobayashi 1990) the lowest transitions from the unresolved A, B valence band-states has been determined at 1.96 eV, the C transition to 2.4 eV. In the MCD spectra there is a positive peak at 1.98 eV. The exact description of the measured spectra could only be achieved by introducing the valence-band structures and the electron–hole exchange energies into the theory. The lowest MCD-signals around 1.98 eV and 2.12 eV have been assigned to the $S_{3/2} \longrightarrow s_e$ and to the $P_{3/2} \longrightarrow p_e$ transitions. The authors propose to attribute these modifications in the MCD spectra to the beginning

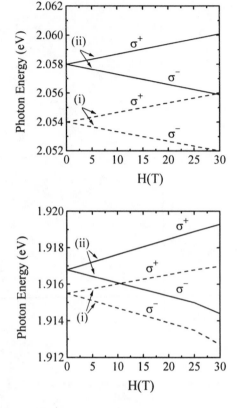

Fig. 7.10. Magnetic field dependence of the energy of the (i) lowest and (ii) second lowest state for σ^+ and σ^- polarization for a CdSe quantum dot with $R = 2.5\,\text{nm}$. From Nomura et al. (1994)

Fig. 7.11. Magnetic field dependence of the energy of the (i) lowest and (ii) second lowest state for σ^+ and σ^- polarization for a CdSe quantum dot with $R = 4\,\text{nm}$. From Nomura et al. (1994)

of the transition from Zeeman splitting to the development of Landau levels.

In a forthcoming work of the same authors (Nomura et al. 1994), a detailed analysis of the magnetic field action on excitons in quantum dots has been presented. There special attention has been paid to the intermediate case of $R \approx \lambda$. In this case, the magnetic-field-induced spin-splitting and deformation of the orbital wave function have to be considered in the same level of approximation. The complete Hamiltonian, comprising potential barrier, Coulomb interaction, valence-band structure and magnetic field has been treated, and in particular, the effect of exchange splitting has been compared with the magnetic field action. The exciton states are obtained by numerical matrix diagonalization.

In Fig. 7.10 and 7.11 the results are shown for the lowest and second lowest state of $R = 2.5\,\text{nm}$ and $R = 4\,\text{nm}$ CdSe quantum dots. These two sizes of quantum dots are representative of the strong and weak magnetic field limit $\hbar\omega_c \ll \Delta E_{\text{exch}}$ and $\hbar\omega_c \gg \Delta E_{\text{exch}}$ with $\hbar\omega_c$ the cyclotron resonance and ΔE_{exch} the exchange energy. In the $R = 2.5\,\text{nm}$ quantum dots the exchange energy splits the lowest pair states into two pairs of states differing in the quantum number of the total angular momentum (see Chap. 3). When the external magnetic field is applied, these levels show the Zeeman splitting and the splitting energy increases linearly with increasing strength of the magnetic field (Fig. 7.10). In the $R = 4\,\text{nm}$ quantum dots (Fig. 7.11), the electron–hole exchange energy can be neglected compared with the magnetic field strength. Here, the obtained deviation from the linear dependence above $15\,\text{T}$ is attributed to the transition from the Zeeman effect to the quantum confined Paschen–Back regime, with a small contribution of the diamagnetic terms in the Hamiltonian.

Finally, it should be mentioned that the magnetic-field induced change in the selection rules has been successfully applied to confirm the exchange splitting experimentally in magnetoluminescence by Nirmal et al. (1995).

7.3 External Fields Acting as Confining Potentials

7.3.1 Magnetic Field Confined Electrons

Over decades, the action of a homogeneous magnetic field on the motion of electrons in three-, two-, or lower-dimensional electron gases has been the subject of continuous interest. In two-dimensional electron gases (2DEG) the condensation in Landau levels has been studied and the Quantum Hall effect discovered. Magnetic quantum dots, however, come into being when the 2DEG is exposed to *inhomogeneous*, perpendicular magnetic fields acting as a magnetic barrier for the motion of electrons. Here we deal with the problem of an electron magnetically confined in an inhomogeneous magnetic field which is zero within a cylinder of radius r_0 and constant outside. We just follow the model presented by Solimany and Kramer (1995) and the details can be found there. Magnetic quantum steps and wells were analytically investigated by Khveshchenko and Meshkov (1993) and Gavazzi et al. (1993).

The Hamiltonian that has been treated by Solimany and Kramer (1995) is

$$\hat{H} = \frac{1}{2m_e}\left(p_r^2 + \frac{p_\phi^2}{r^2} + p_z^2\right) \tag{7.7}$$

for coordinates r within the cylinder $r < r_0$ and

$$\hat{H} = \frac{1}{2m_e}\left(\boldsymbol{p} - \frac{e}{c}\boldsymbol{A}\right)^2 \tag{7.8}$$

outside the cylinder $r > r_0$. Here, m_e is the mass of the electron. The magnetic field is described in cylindrical coordinates by

$$B(r) = \begin{cases} B & \text{for} \quad r > r_0 \\ 0 & \text{for} \quad r < r_0 \end{cases}. \tag{7.9}$$

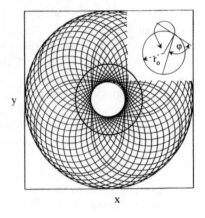

Fig. 7.12. Classical orbit of an electron in a magnetic quantum dot with radius r_0. The inset shows the derivation of condition for closed orbits. From Solimany and Kramer (1995)

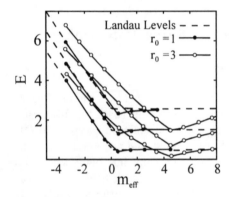

Fig. 7.13. Energy eigenvalues E in dependence on the effective angular momentum $m_{\text{eff}} = m + r_0^2/2$ for the cases $r_0 = 1$ and $r_0 = 3$. The dashed lines show the Landau levels ($r_0 = 0$). From Solimany and Kramer (1995)

The electron moves in an (r, ϕ) plane and the magnetic field acts perpendicularly to that plane in z direction. Figure 7.12 shows one of the numerically calculated trajectories obtained for the movement of the electron within the potential (7.9) The condition for closed, periodic orbits (corresponding to the quantum mechanical eigenstates) has been derived in analogy to the condition of closed circles in conventional billiards (reflection angle equals multiples of π).

The requirement that the wave function ψ has to vanish for $r \to \infty$ and for continuity of ψ and its derivatives at the boundary $r = r_0$, yields the energy eigenvalues E which are plotted for two values of r_0 in relation to the angular momentum m in Fig. 7.13. Additionally, the case $B = \text{const.}$ (homogeneous magnetic field) is shown. Comparing the result for inhomogeneous and homogeneous field conditions, we see that the eigenenergies decrease for $m > 0$ with respect to the Landau levels and an increase for $m < 0$. The electron is localized within the region of the dot, however, it needs a lower (higher) energy for the motion with positive (negative) angular momentum,

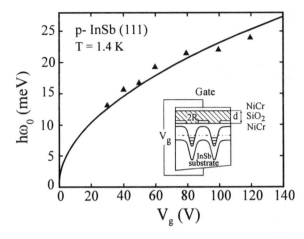

Fig. 7.14. Quantization energy $\hbar\omega_0$ of field-effect confined InSb quantum dots as a function of gate voltage. The solid line is calculated with assumption of nearly parabolic potential barrier. From Merkt (1993)

than in a homogeneous magnetic field. The lowest eigenvalue corresponds to $m_{\text{eff}} = r_0^2/2$, i.e., $m = 0$.

7.3.2 Electric Field Confined Electrons and Transport Properties

A further kind of quantum dot originates from the lateral confinement of a two-dimensional electron gas (2DEG) in a semiconductor heterostructure when a gate voltage of special geometry is applied. The field-effect confined dot electrons are surrounded by reservoirs of electrons of the 2DEG (the 'matrix' is conducting!) and interesting transport properties are expected. In these quantum confined systems, direct information about the confinement is accessible by probing the intraband transitions between the confined electron states via far-infrared spectroscopy. An example of a quantum structure of this type is based on InSb (Merkt 1993). The semiconductor InSb has a very small effective conduction-band mass of $m_{\text{eff}} = 0.014m_0$ and thus the advantage of comparatively large energy spacings between the confined electron levels up to 25 meV (typical quantization energies of 10 meV correspond to wavelengths of 100 μm in the far-infrared). The realization of an InSb quantum dot is sketched in Fig. 7.14. It consists of a metal-oxide-semiconductor structure with the alloy NiCr condensed on InSb as a Schottky barrier. The inset of Fig. 7.14 shows a schematic picture of the band structure across the dot if a strong negative gate voltage is applied. The Fermi energy E_{F} is pinned at the NiCr/InSb interface above the valence-band edge. In the areas between the NiCr contacts, mobile inversion electrons are induced when the gate voltage reaches strong negative values.

The simplest approximation of the lateral confining potential resulting from the voltage applied is the parabolic potential barrier. The single electron properties then are those of the harmonic oscillator. A more sophisticated model is that of a conducting plate with a circular hole of radius R (idealized structure without charged interfaces and no discontinuity in the dielectric constants) treated by Junker et al. (1994). These models predict an increase in the single-electron energy $\hbar\omega_0$ proportional to the square root of the gate voltage. The experimental observation of this functional dependence between the confined electron energy $\hbar\omega_0$ and the gate voltage V_g only succeeded in the case of ideal dot structures. Most often technological reasons like the properties of the Schottky contact and the differences in the dielectric constants cause substantial deviations of the potential barrier from the parabolic shape. Moreover, electron–electron interactions, nonparabolicity of the band structure, intrinsic fields at the semiconductor–oxide interface, etc., can result in weaker shifts of the resonance frequency with increasing gate voltage. However, in selected samples, the tuning of the frequency ω_0 with gate voltage could be observed as shown in Fig. 7.14 (Junker et al. 1994; Merkt et al. 1993). A further example of electric-field-induced dot structures has been given, for example by Demel et al. (1990) developing deep-mesa etched quantum dots from an AlGaAs/GaAs heterostructure. More detailed information about spectroscopic experiments to determine the resonance energy as a function of gate voltages can be obtained in the reviews given by Heitmann (1995), Merkt (1993), Chakraborty (1992) and the references therein. Often additional information has been derived by investigating the Zeeman splittings of the confined electron levels under the action of an external magnetic field.

The importance of the structures discussed above lies in their transport properties and potential for single-electron electronics (bit-by-bit processing). There are excellent reviews about studies of transport in systems where the quantum dots are defined by electrostatic depletion of parts of the sample using special designed gates (Haug 1994; Kouwenhoven et al. 1991; Beenakker and Van Houten 1991; Chakraborty 1992; Van Houten 1992; Kastner 1992; McEuen et al. 1993; and Bryant 1993). In these experiments, transport is influenced not only by the quantum mechanical confinement but also by the charging energy since the small capacitance of the dot structure plays a role. The Coulomb charging energy E_C is the energy that is required to add one additional electron into the dot,

$$E_C = \frac{e^2}{2C}. \tag{7.10}$$

The charge of a single electron becomes important if the charging energy E_C exceeds the thermal energy $k_B T$. In this case the transport through the dot is not a continuous flow of charges but a discrete transport of single electrons. If an additional gate is capacitively coupled to the dot and used to modify the electrostatic energy of the system (change from blocking to tunneling conditions), these structures can be considered as single-electron tunneling transistors.

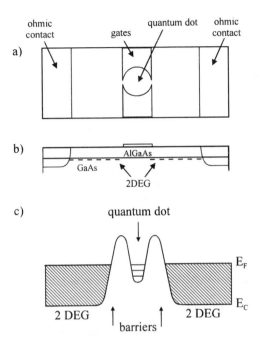

Fig. 7.15. Scheme of a gated quantum dot structure working as a single-electron transistor: (**a**) top view, (**b**) side view, (**c**) band structure along the current direction. According to Haug (1994)

Such a structure realized in an AlGaAs/GaAs heterostructure is schematically shown in Fig. 7.15 (Haug 1994). The 2DEG serves as drain and source of the transistor. On the top of the AlGaAs-layer a capacitively coupled gate determines the size and the charge of the quantum dot. Large negative voltages deplete the 2DEG underneath the gate structure and an isolated quantum dot is formed in the middle. Through the capacitively coupled gate, the electrostatic potential of the enclosed electronic system can be changed.

By the change in the gate voltage, the charge on the quantum dot can be increased and for $E_C \gg k_B T$ two typical situations will appear: Coulomb blockade of the transport and single electron tunneling, schematically illustrated in Fig. 7.16 (Kouwenhoven et al. 1991). Adding an electron to the dot changes the electrochemical potential of the dot by $\mu_e(N+1) - \mu_e(N)$. This difference can become large for large energy splittings of the quantum dot states and/or small capacitance of the dot. As seen schematically in Fig. 7.16a, this gap in the chemical potentials can lead to a blockade of tunneling of electrons into and out of the dot, known as the Coulomb blockade. Changing the gate voltage can eliminate the Coulomb blockade (Fig. 7.16b) by adjusting $\mu_e(N+1)$ between μ_l and μ_r and now an electron can tunnel through the dot. The tunneling current exhibits characteristic oscillations, known as Coulomb oscillations (see Beenakker 1991 and references therein).

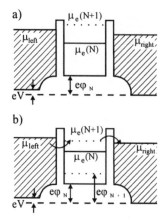

Fig. 7.16. Principle of sequential electron transport through a quantum dot: **(a)** Coulomb blockade of electron tunneling because $\mu_e(N) < \mu_l, \mu_r < \mu_e(N+1)$, and **(b)** single-electron tunneling for $\mu_l > \mu_e(N+1) > \mu_r$. V denotes the small voltage across the sample [$eV \ll \mu_e(N+1) - \mu_e(N)$], ϕ_N is the electrostatic potential with N electrons in the dot with discrete energy levels E_N, μ_l and μ_r are the chemical potentials of the left and right reservoirs of the 2DEG, and μ_e is the chemical potential of the dot. According to Kouwenhoven et al. (1991)

The present research work on transport in quantum dots concentrates on the treatment of the full many-body problem including Coulomb interaction. First attempts in this direction have been published, for example, by McEuen et al. (1993), Tanaka and Akera (1996), and Hallam et al. (1996). Transport through antidot structures is described by Rotter et al. (1995) and the references therein. Selection rules for transport processes of few-electrons quantum dots can be found in the work of Pfannkuche and Ulloa (1995).

8. Nanocrystals of III–V Compounds

Growth, preparation and structural characterization of quantum structures based on III–V semiconductor compounds have been described in Chap. 2. Here we report on the connection between three-dimensional nanostructuring and observation of confinement-induced energy shifts in the spectra of photoluminescence, photoluminescence excitation and absorption of III–V materials. In contrast to II–VI quantum dots in glasses, most of the III–V quantum dots do not occur as spherical entities. In particular, if their preparation is based on epitaxial deposition methods, they may show a great variety of different shapes. Hence these structures exhibit very different confining potentials. It is the aim of this chapter to test how far a correlation between shape and electronic states can be proven in the optical properties as a sign of three-dimensional quantum confinement. The chapter provides only a limited part of the presently known data and we refer to the references in the presented papers themselves.

8.1 Spherical Quantum Dots in Polymers and Glasses

Quantum crystallites of III–V semiconductors, in particular GaAs, rarely exist as spherical inclusions inside glasses or other matrices. Reasons for this behavior are numerous difficulties during their preparation, such as the high toxic reagents or the fast oxidation at usual glass technology temperatures. The few known examples concern the III–V materials GaAs and InP. Turning first to the near-infrared material GaAs, the published experimental data show the common result of an enormous confinement-induced energy shift of $\sim 1\,\mathrm{eV}$ up to the visible spectral range. For example, Olshavsky et al. (1990) succeeded in precipitating $\sim 4\,\mathrm{nm}$ GaAs microcrystals in polar organic solvents. The quinoline-soluble particles have been studied by TEM. Slightly prolate GaAs nanocrystals with an average major axis of $4.5\,\mathrm{nm}$ and minor axes of $3.5\,\mathrm{nm}$ ($\pm 10\,\%$) were observed. These GaAs particles show a well-structured absorption peak at $480\,\mathrm{nm}$ (!). Luong and Borelli (1989) and Wang and Herron (1991b) have shown that porous glass can be impregnated with GaAs particles. This method has been further developed more recently by Justus et al. (1992) and Hendershot et al. (1993) who use the porous Vycor glass (Corning, Inc.) and incorporate GaAs and InP into the microscopic voids to produce quantum confined structures. The GaAs nanoparticles are fabricated in pores with a nominal size of 4 nm. The linear absorption spectra exhibit broad featureless absorption with a tail at 700 nm. For this new material the nonlinear-optical properties have been studied and a two-photon absorption coefficient of 30 cm/GW has been found as well as the nonlinear refractive index of -5.6×10^{-12} (at 1060 nm). Whereas the two-photon absorption efficiency is similar to bulk GaAs, the nonlinear refractive index of the GaAs impregnated glass is enhanced by more than one order of magnitude. Another approach to prepare uniformly sized GaAs quantum dots has been proposed by Salata et al. (1994). The authors have developed a gas-aerosol reactive electrostatic deposition of semiconductor nanocrystals in a polymer matrix (for example PVA). The average size of the GaAs quantum dots is 13.4 nm with a polydispersity of less than 10 %. The absorption onset is blue-shifted by 40 nm with respect to the bulk. A discrete structure owing to the lowest one-pair transition can be seen at 795 nm. The developed method provides a high monodispersity of the nanoparticles.

The precipitation of InP in porous Vycor glass with 4 and 15 nm pore size has been reported by Hendershot et al. (1993). The absorption spectrum of the glass containing the smaller crystallites has its onset at 750 nm and achieves the maximum at 500 nm. The 15 nm InP particles display a smoother absorption tail between 1 µm and 700 nm due to the more inhomogeneous broadening and defects. Giessen et al. (1996c) identified the first excited pair states for InP quantum dots by measuring selective absorption bleaching using fs-pump-probe spectroscopy at room temperature.

8.2 Quantum Dots Obtained by Deep-Etching and Interdiffusion

Lithography-based techniques offer the possibility to fabricate lateral, *ordered arrays* of nanostructures of any degree of confinement and thus go beyond the conventional techniques which provide statistically distributed structures. Optimum technology allows the formation of quantum dot arrays down to 30 nm dot diameter (Steffen et al. 1996). The etched dots are more cylindrical, but the etching techniques allow both single dot spectroscopy as well as the spectroscopy of the whole dot array. The first problem dealt with was the examination of the predicted enhancement of luminescence intensity with decreasing size, based on arguments of concentrated oscillator strength and superradiance. Corresponding studies of radiative recombination and emission has been performed at etched structures derived from $GaAs/Al_xGa_{1-x}As$ quantum wells (Wang et al. 1992a). However, most papers on etched structures did not report an enhancement but, on the contrary, a decrease in the luminescence efficiency with decreasing structure size (Forchel 1988; Wang et al. 1992a). To explain this behavior, a slowing down of the energy relaxation has been suggested for dot sizes below 100 nm resulting from decreased phonon scattering rates (phonon bottleneck). An additional extrinsic effect, the fast carrier diffusion to nonradiative recombination centers at the surface, is typically found in etched structures. By Steffen et al. (1996) the confinement effect has been investigated for cylindrical dots derived from a 5 nm $In_{0.1}Ga_{0.9}As$/GaAs single quantum well structure by low-voltage electron beam lithography and wet chemical etching. The maximum observed energy shift of 14 meV for a 27 nm dot diameter has been compared with theoretical calculation using cylindrical boxes. Good agreement was achieved. With similar structures the enhancement of spin splitting owing to the spatial confinement could be proven (Bayer et al. 1995) applying an external magnetic field up to 8 T. The increase of the splitting energy with decreasing dot diameter has been attributed to the enhancement of the g factor of the hole ground state, which obtains admixtures from hole states with nonzero orbital angular momentum.

Starting from III–V quantum wells with high-quality, sharp interfaces the formation of quantum dot structures is possible by spatially localized, thermally activated interdiffusion. By this local interdiffusion, a lateral pattern inside the quantum well can be formed and buried quantum structures can be defined (see for example Forchel 1988). By Brunner et al. (1992) quantum dots of high quality have been fabricated by laser-induced thermal interdiffusion of a GaAs/AlGaAs quantum well structure. The dot dimensions are defined by a square frame of width between 1000 nm and 200 nm. The spectra show sharp transitions from a single quantum dot. When the dot width is reduced down to 450 nm, pronounced peak structures and a blue shift of about 10 meV are observed. The blue shift and the energy splitting of the luminescence spectrum can well be described by a model based on parabolic potential well and neglecting Coulomb interaction and interband coupling. This model results in equidistant energy levels for the confined electron and hole states. For a dot width of 450 nm, the quantization energy is 6 meV for the electron states and 4 meV for the heavy hole states as observed in the experiment.

8.3 Quantum Dots due to Spatially Isolated Potential Fluctuations

In the preceding chapters, zero-dimensionality of semiconductor structures was discussed by considering an ensemble of isolated, noninteracting quantum dots with nearly identical energies. This limit (a) is appropriate to describe II–VI quantum dots embedded in transparent matrices or InAs islands obtained by self-organized growth on GaAs substrates (see the section below). However, the spatial 3D-periodicity of the electron and hole-wave functions can likewise be destroyed by naturally given, local fluctuations in the material structure, such as width fluctuations in quantum wells or perturbations of

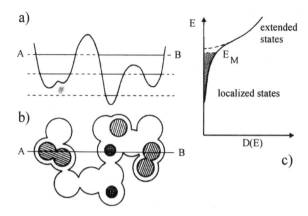

Fig. 8.1. Localization of electronic states within a landscape of randomly distributed local potential minima; (a) potential fluctuation versus a spatial coordinate; (b) contour lines of equal potentials; (c) density of states (E_M denotes the mobility edge). The line AB in (a) is the cut between A and B, drawn in (b)

the corrugation in superlattices grown on faceted substrate surfaces.[1] Therefore, to describe epitaxially grown structures, a second way of modelling can become relevant: the limit (b) of randomly distributed, local potentials localizing excitons and forming single, individual quantum dots of different energies (similar to deep localized states at composition fluctuations in mixed crystals). This case represents the extension of the conventional model of localized states towards zero-dimensionality. Discrete energies and δ-like density of states are characteristic for deep localized states within a potential landscape. These states are no more able to tunnel to the next potential minimum during their lifetime at given temperature and form the exponential tail in the density of states. They appear in the spectra red-shifted with respect to the energy of the corresponding extended states. The principle of localization is illustrated in Fig. 8.1. From this figure we see that the quantum dot model treated so far is obtained, when the different potential minima are decoupled by thick barriers and become almost identically.

The both limits exhibit typical optical properties and can be distinguished in optical experiments. Typically for case (a) is the observation of strictly size-correlated peak shifts to higher energies compared to the bulk when decreasing the dimensions. In time-resolved experiments, we expect uniform decay of a predominantly homogeneously broadened line. With increasing intensity, subsequent filling of the excited states results in the appearance of high-energy peaks at energy positions defined by the confinement. The case (b) predominates if we observe a strong inhomogeneously broadened luminescence line, a high sensitivity of the line shape with respect to temperature change (thermal activated hopping may even result in blue shifts of the line maximum with increasing temperatures), and a nonuniform decay behavior within the spectrum due to the motion of excitons between the different po-

---

[1] Substrates with orientation of the surface in direction of high-crystallographic index, like $(n11)$, exhibit a naturally given microroughness on atomic scale.

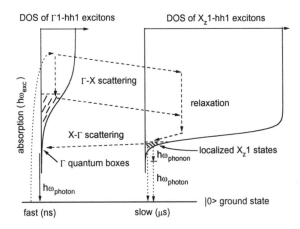

Fig. 8.2. Relaxation and recombination in type-II-[311]-GaAs/AlAs superlattices. Strongly localized states in the tails of density of state (DOS) have been considered as quantum boxes. From Langbein et al. (1996)

tential minima. The relaxation dynamics of excited states is influenced by scattering processes with the continuum of extended states.

Experimental evidence for the formation of three-dimensionally confined states has been obtained by spatially-resolved detection of extremely narrow luminescence lines in structures that have such local potential fluctuations. The isolation of luminescence from a single quantum dot requires special arrangements since the density of states in these quantum 'dots' is small compared with that of the continuum states. One possibility is the resonant electron injection from the X valley to the Γ point in type-II quantum structures as illustrated in Fig. 8.2 (see e.g Langbein et al. 1996, and the references therein). After optical excitation at the Γ point, the electrons undergo an ultrafast scattering to the X-minimum. The long lifetime in the X-subband minimum allows a scattering process back to the localized states at the center of the Brillouin zone (Γ-quantum boxes). The X-band minimum acts as the charge reservoir for the injection of nearly resonant electrons from the X point to localized states at the Γ point. The efficiency of the scattering depends on the energy separation between the Γ and X points, on the amount of localization energy (localization depth), the mixing effect between Γ and X wave functions and on the strength of electron–phonon interaction.

One of the first concepts for naturally given quantum dots in quantum wells was put forward by Zrenner et al. (1994). These authors consider an extremely narrow quantum well as a disordered array of quantum dots with the local potential fluctuations representing the potential for the zero-dimensional confinement. A variety of localized states of arbitrary energy are assumed to be created below the ground state exciton at the Γ point. The authors take their arguments for this concept from the occurrence of very narrow lines in luminescence energetically below the fundamental absorption edge. The quantum dots below the Γ minimum become resonant to the X minimum by tuning the X-subband energy via an external electric field.

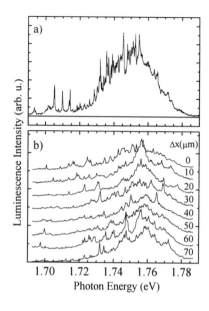

Fig. 8.3. Photoluminescence of a (9ML GaAs/14ML AlAs) type-II-[311] super-lattice, excited with $I_{exc} = 4\,W/cm^2$ at 1.96 eV. (a) detection spot 6 μm diameter, $T = 30\,K$, (b) detection spot 6 μm diameter, but changing its lateral position across the sample, $T = 20\,K$. From Langbein et al. (1996)

Gammon et al. (1996) studied the luminescence of excitons localized at interface fluctuations in a narrow GaAs/AlGaAs quantum well by high spatial resolution, and the excited state spectrum of single GaAs quantum dots was revealed. The peaks are extremely narrow and exhibit a fine structure splitting.

The lateral growth process at [n11] surfaces has been described, for example, by Nötzel et al. (1993, 1994a). Quantum box-like structures form spontaneously during self-organized growth on GaAs(311) substrates. The corrugation height amounts to ~ 0.3 nm (Wassermeier et al. 1995). The model of Zrenner et al. (1994) has been applied to corrugated superlattice structures by Xu et al. (1995). The deeper localized states are produced by perturbation of the periodic corrugation of the interface. They lie energetically resonant to the X minimum. The strong localization effect in GaAs/AlAs corrugated superlattices grown in (311)-oriented GaAs substrates has been demonstrated by temperature-dependent measurements of the luminescence and observation of a red-shift of the exciton emission relative to the emission energy of a (100) reference sample.

Single dots in GaAs/AlAs corrugated superlattices have been spatially resolved as narrow peaks in photoluminescence by Langbein et al. (1996) (Fig. 8.3). Measuring the degree of polarization of luminescence, the optical anisotropy could be determined. By comparison with calculations, the aspect ratio of the GaAs quantum box lengths in the (1$\bar{1}$0) and (113) directions could be derived. The narrowness of the luminescence lines is surprising and the line-width seems to correspond to the pure quantum-mechanical lifetime. The recombination spectrum consists of a series of sharp lines and the spectral width of the spectrum reflects the distribution in energies due to statistical fluctuations in size (inhomogeneous line-broadening with $\Gamma_{\text{hom}} \ll \Gamma_{\text{inhom}}$).

8.4 Quantum Dots Resulting from Self-organized Epitaxial Growth

The specific property of quantum dots grown in a self-organized mode (see Sect. 2.5) consists in a steady change of the eigenenergies of the nanostructured semiconductors not only owing to three-dimensional confinement, but also because of the variations of strain and shape. The growth mechanism mostly does not allow a continuous variation of the size of the nanostructures. The best investigated system at present are InAs islands grown on GaAs (lattice mismatch 7 %). Recently also the realization of InP quantum dots grown in $Ga_{0.5}In_{0.5}P$ has been reported by Petterson et al. (1996). The spatial extension of the grown islands is predominantly controlled via the strain, for example, via changing the x composition of $In_xGa_{1-x}As$ mixed crystals. The determination of the exact x composition, however, is a puzzle. In particular, the actual amount of indium, incorporated into the dots, can differ owing to diffusion of indium during the island formation. The wetting layer itself gives an efficient luminescence signal close to the energy of the quantum dots [for 1.7 monolayer (ML) InAs the luminescence has been reported at $\sim$1.38 eV by Grundmann et al. (1995a)]. A few examples of realized structures are given in Table 8.1. There we try to summarize published data with clear identification of both the grown geometry as well as the corresponding peaks in photoluminescence (PL).

To our knowledge, to date two theoretical works have appeared which deal with the Schrödinger equation for the specific confining potentials of pyramidal quantum dots under strain. Marzin and Bastard (1994b) approximated the pyramids by cones lying on the 2D wetting layer with identical strain in the dots and the wetting layer. The numerical approach takes into account modification of the potential owing to In segregation when the island is overgrown with the GaAs cap layer. A thorough study of the strain distribution in and around the InAs pyramidal quantum dots and the consequences for the electronic states and phonon energies has been performed by Grundmann et al. (1995b). The Coulomb energy has been treated in perturbation theory. If the shape of the dots does not change during the growth, the strain distribution gives a size-independent correction. For the LO-phonon energy a strain-induced energy change of $\Delta E = 2.2$ meV has been calculated, giving $E_{\text{LO}}^{\text{QD}} = 32.1$ meV. In contrast to Marzin and Bastard (1994b), the authors use locally varying, anisotropic effective masses in the Hamiltonian for calculation. In both papers the electron levels fall into the continuum of the wetting layer for base lengths of the pyramids below $\approx$ 6 nm. The holes may have several excited states. For the energies of the first

Table 8.1. Ground state energy of InAs-based self-organized quantum dots grown on (100) GaAs substrates and determined by measuring the photoluminescence (PL) at low temperatures and nonresonant excitation

Island material	Barrier material	Geometry	Maximum of PL	Reference
InAs	GaAs within a GaAs/ $Al_{0.3}Ga_{0.7}As$-SL	pyramids base 12±1 nm height 5±1 nm	1.1 eV	Grundmann et al. (1995a)
$In_{0.5}Ga_{0.5}As$	GaAs	diameter 25 nm height ≤2.5 nm	1.27 eV	Fafard et al. (1994b)
InAs	GaAs	pyramids base 19 nm height 2.1 nm	1.28 eV	Marzin et al. (1994a)
InAs	GaAs	diameter 20 nm height 6 ML	1.35 eV	Lubyshev et al. (1996)
InAs	$Al_{0.35}Ga_{0.65}As$	diameter 24.4 nm height 4.4 nm	1.46 eV	Leon et al. (1995)
InAs	GaAs slightly misoriented	diameter 7 nm at terraced surface	1.468 eV	Li et al. (1994)
InAs	AlAs	diameter 16.2 nm height 2.9 nm	1.77 eV	Leon et al. (1995)
$Al_{0.45}In_{0.55}As$	$Al_{0.35}Ga_{0.65}As$	diameter 17 nm height 3 nm	1.88 eV	Fafard et al. (1994a)
$In_{0.55}Al_{0.45}As$	AlAs	diameter 23 nm height 4.1 nm	1.92 eV	Leon et al. (1995)
$In_{0.67}Ga_{0.33}As$ $In_{0.5}Ga_{0.5}As$ $In_{0.5}Ga_{0.5}As$	GaAs GaAs GaAs	diameter 15 nm diameter 13 nm diameter 36.5 nm	1.276 eV 1.282 eV 1.1 eV	Farfad et al. (1995)

excited hole states values close to multiples of the LO-phonon energies have been found. Figure 8.4 shows the result of the calculation made by Grundmann et al. (1995b) for the transition energy between the electron and hole ground state as a function of dot size. For an InAs pyramid with square base of 12 nm grown on 1.7 ML wetting layer, the lowest optical transition is obtained at an energy of 1.083 eV. This value is in good agreement with experimental data published by the same group (Grundmann et al. 1995a, Table 8.1). However, the data reported by, for example, Lubyshev et al. (1996) are difficult to explain within the pyramid-based model. Because the strain con-

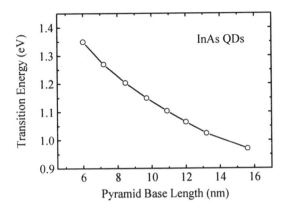

Fig. 8.4. Transition energy between the electron and hole ground state as a function of dot size. From Grundmann et al. (1995b)

tribution in the Hamiltonian changes when the island shape, the substrate properties or the composition x in mixed crystals are varied, the solution of the Schrödinger equation is different for every shape and x of the grown islands and a general pattern is difficult to observe. The system presently investigated the best is the pyramidal-shaped InAs/GaAs quantum dot described by Grundmann et al. (1995a). For this structure both the excited hole states as well as the multiphonon-assisted relaxation process were revealed in the experiments (Grundmann et al. 1996; Heitz et al. 1996). First hints at excited hole states have also been found in photoluminescence excitation spectra of $In_{0.5}Ga_{0.5}As/GaAs$ quantum dots and by means of photoluminescence with tunable excitation (Fafard 1994b; 1995). The photoluminescence spectra show a strong enhancement in intensity which is correlated with the resonance condition between interlevel spacings and phonon energies. Quantum structures with smaller interlevel spacings display strong level filling and emission from the excited states.

As already detailed in Chap. 5, the search for the phonon contribution to the relaxation processes (phonon bottleneck) is of particular interest here. A rapid energy relaxation pathway by multiphonon-assisted tunneling has been proposed by Sercel (1995), mediated through a deep-level trap coupled to the energy states of the quantum dot. Tunneling rates of $> 10^{10}$ s^{-1} have been calculated for a 5 nm radius $In_{0.5}Ga_{0.5}As/GaAs$ quantum dot where the deep trap has been assumed to be close to a sphere within a volume of ~ 10 nm radius centered on the quantum dot. The author suggests that the formation of trap states in the interface region of the quantum dot during the growth process may possibly explain both the high efficiency of luminescence and the difficulty to observe a significant phonon bottleneck in $In_xGa_{1-x}As$ quantum dot structures.

Finally, it should be mentioned that quantum dots grown in a self-organized way moreover allow us to realize type-II confinement (indirect-gap quantum dots, see Chap. 2). An example has been given by Hatami et al.

(1995) with the GaSb/GaAs system. This system shows similar values for the lattice-mismatch as InAs/GaAs and thus the formation of three-dimensional GaSb islands is easily achieved after deposition of a few monolayers. The dots, investigated by Hatami et al. (1995) exhibit a rectangular base shape and a lateral size of $\sim 20\,\mathrm{nm}$. The band alignment scheme of a GaSb/GaAs 2D heterostructure has the typical staggered behavior as known for type-II heterojunctions. The formation of GaSb quantum dots is revealed in the photoluminescence spectra by a broad peak shifted to lower energies compared to the wetting layer luminescence. In this structure the luminescence originates from radiative recombination of confined holes and electrons spatially separated but mutually attracted via Coulomb interaction.

A similar, indirect-type transition related to Γ-X crossing in k space can be observed in InAs/GaAs quantum dots when increasing the externally applied pressure. This has been shown by Li et al. (1994) for InAs/GaAs quantum dots at $> 4.2\,\mathrm{GPa}$. The quantum dot luminescence from the electron and heavy hole states at the Γ point vanishes with increasing pressure, and besides the type-I transition within the InAs (electron at X_{xy} and heavy hole), a type-II transition arises from X states in GaAs and the heavy hole states in InAs.

8.5 Stressor-Induced Quantum Dots

Strain-confinement can be produced by putting a so-called stressor on the top of a quantum well. This stressor consists of a semiconductor material with a strongly different lattice constant. The stressor deforms the quantum well laterally and generates a (mostly parabolic) confinement potential. Zhang et al. (1995) used carbon dots on top of GaAs/GaAlAs quantum wells for the stressors. The authors propose that strain patterning enhances the luminescence efficiency over that of the equivalent quantum well. The reason is that, since dimensionality is reduced from 2D to 0D, nonradiative recombination is effectively suppressed. The electron–hole pairs are localized in the strain-induced potentials and only temperature-dependent activation allows the transfer of excitons above the barrriers. The confinement reduces the exciton mobility and thereby the probability of reaching nonradiative recombination centers. This reduction of nonradiative recombination could become interesting for possible applications of these quantum structures.

Likewise, it has been shown by Lipsanen et al. (1995) that the lateral confinement of quantum dots by growing InP stressors on InGaAs/GaAs quantum wells yields structures with high luminescence efficiency. For higher excitation densities even the excited states become observable. Their equidistant energy separation can well be fitted within the harmonic oscillator model, assuming essentially parabolic potential barriers (Tulkki and Heinämäki 1995).

9. Nanocrystals of Indirect-Gap Materials

In indirect-gap semiconductors the energetically lowest electronic transition is optically forbidden and appears in the spectra only by assistance of momentum conserving phonons or impurities. The radiative recombination of electron–hole pairs has a very low probability. When making these crystals finite, the spatial confinement is correlated with a large extension of the pair state in momentum space and one expects a sharp increase in the oscillator strength. Some of these semiconductors like Si or Ge are well-established materials in microelectronics, and the quantum confinement might open an exciting perspective in producing both the energy shift of the light emission into the visible range and the increase in its efficiency.

The intent of this chapter is to provide an overview of the evidence of quantum confinement in nanostructures made from indirect-gap semiconductors. In particular, we deal with the most intensively investigated material, the nanostructured silicon. It is not possible to mention all of the tremendous activities on research of nanocrystalline and porous silicon, and interested readers are directed to the reviews by Smith and Kollins (1992), Lockwood (1994), Feng and Tsu (1994), Brus (1994), and Brus et al. (1995) with the references cited therein. Other nanostructures of indirect-gap semiconductors, like Ge and AgBr, have been described, for example, by Hayashi et al. (1990), Masumoto et al. (1992), Marchetti et al. (1993), and Maeda (1995). Even if the first results of the optical studies suggest that the indirect-gap character of the semiconductors also survives in very small particles, the understanding of the radiative and nonradiative processes remains in need of further improvement.

9.1 Theoretical Description

Quantum confinement effects appear when the geometric dimensions of the nanostructures reach the value of the bulk excitonic Bohr radius a_B. In silicon, $a_B \sim 5\,\text{nm}$ holds for the free exciton. To test the actual chance for observing quantum confinement in indirect-gap semiconductors, theoretical calculations have been performed based on both effective mass approximation (EMA) (Takagahara and Takeda 1992; Xie et al. 1994), and tight binding methods (Proot et al. 1992; Delerue et al. 1993; Huaxiang et al. 1993).

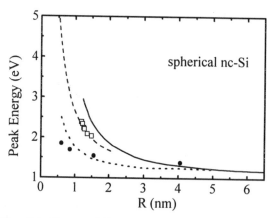

Fig. 9.1. Energy shift of the absorption edge in dependence on particle radius for Si quantum dots. Solid line: EMA calculation from Takagahara and Takeda (1992), dashed line: LCAO calculation from Delerue et al. (1993), dotted line: calculation from Hill and Whaley (1995a), squares: experimental data from Furukawa and Miyasata (1988), dots: luminescence data from Schuppler et al. (1994)

The theoretical analysis in the frame of EMA starts immediately with the full treatment according to the Hamiltonian (3.25) using the Luttinger term (3.26) for the valence band (see Chap. 3). Takagahara and Takeda (1992) solved (3.25) for Si quantum dots assuming infinite barrier potentials and taking into account the anisotropy of the conduction band. The size-dependence of the absorption edge thus obtained is shown in Fig. 9.1 (solid line). The computations cover the range of sizes between $R = 1.2$ nm and 7 nm. The calculated transition probability of the radiative recombination derived from the overlap of the wave functions grows drastically over more than 9 orders of magnitude with decreasing size of the quantum dots. Further improvements of the theoretical description take into account phonon-assisted transitions (Hybertsen et al. 1994; Xie et al. 1994; Suemoto et al. 1993, 1994; Calcott et al. 1993b), as well as the effective medium character of the sample owing to the high volume fraction of semiconductor material (Xie et al. 1994; Lannoo et al. 1995).

In the range of very small sizes the tight binding approach seems to be more appropriate. Applying the LCAO method (linear combination of atomic orbitals), Delerue et al. (1993) calculated the electronic structure of silicon quantum wires and dots. The highest confinement, i.e., the strongest blue shift of the absorption edge, was obtained for the spherical shape of the crystallites. The data obtained for the cylindrical wires yields the energy states in the visible range of the spectrum only for very small radii below 1 nm. The results for the spherical nanocrystals are likewise plotted in Fig. 9.1 covering sizes between $R = 0.5$ nm and 2.5 nm (dashed line). The curve obeys a $d^{(-1.39)}$ law, where d is the diameter of the spheres. The value of the exponent is smaller than 2, the value which corresponds to the exponent of the particle-in-a-spherical-box solution in EMA. Therefore the authors point out that the bands cannot be described in simple parabolic approximation, at least at very small sizes. For larger values of d, however, the tendency towards the exponent 2 could be derived from the analysis of the LCAO results. Huaxiang et al. (1993) presented a similar tight binding model. The electronic structure, the momentum matrix element and the absorption spectra were calculated for small spherical Si clusters with diameters between 1 and 2 nm. Compared with the bulk crystalline Si, a new absorption peak in the visible range of the spectra was found, shifting with decreasing size to higher energies and showing the growth in oscillator strength by its increasing intensity.

The third theoretical curve in Fig. 9.1 (dotted line) shows the result of a calculation of the confined pair states by treating the Coulomb interaction nonperturbatively in a time-dependent tight-binding technique which was performed by Hill and Whaley (1995a). The exciton energies predicted by that method are considerably smaller. In this calculation, the Si particles have been modelled as so-called H-terminated nanocrystals, i.e., the surface dangling bond orbitals are saturated by hydrogen atoms. It has been shown that the inclusion of the Coulomb interaction becomes essential when calculating the charge distributions of the electron and hole wave functions.

In Fig. 9.1, besides the three representative results of theory, some experimental data are also plotted for comparison. Actually, the comparison with experimental data is not a priori the proof for quantum confinement. To avoid ambiguity, we have chosen data obtained from Si nanocrystals and not from porous Si, because the latter often contains a mixture of more wire-like and more dot-like structures and the determination of their dimensions is very difficult. Therefore, for Si nanocrystals both the values from the absorption (Furukawa and Miyasato 1988) and the data of luminescence (Schuppler et al. 1994) are plotted for comparison in Fig. 9.1. Good coincidence exists between the data of Furukawa and Miyasato (1988) and the calculations in the range between 1 to 3 nm radius. The photoluminescence data of Schuppler et al. (1994) lie at somewhat lower energies, but are in good agreement with the calculations of Hill and Whaley (1995a). The coincidence in the latter case is given by the structure of the nanocrystals. The photoluminescence data are obtained from well-defined spherical, surface-passivated (by hydrogen) Si nanocrystals, i.e., under conditions as modelled in the theory.

Delerue et al. (1993) also examined the size dependence of the radiative transition probability. The transition matrix element evaluated is very sensitive to the symmetry and the k-space overlap of the electron and hole wave functions. The general tendency is an enhancement of about 5 orders of magnitude. The decay time predicted by the theory reaches the range of microseconds for sizes below $d < 2.5$ nm. At that size the Si nanospheres are characterized by optical properties between those of direct and indirect gap semiconductors. This theoretical result, however, does not automatically imply an increased quantum efficiency resulting from confinement. As we will see below, the luminescence decay time and the quantum yield are determined by the ratio between radiative and nonradiative transitions which can change for reasons other than quantum confinement.

As a further interesting result, the theoretical determination of the absorption spectrum both by Delerue et al. (1993) and Huaxiang et al. (1993) reveals a strong increase in the energy range around 3–4 eV, probably arising from the direct optical transition known from bulk crystalline Si. Even if the calculations determine the optical threshold in relation to size in the range between 1.5 and 2.0 eV, the strongest absorption occurs in the UV region. Although the absorption near the band edge is already rather efficient and the subject of a blue shift with decreasing size, the absorption maximum in the UV exceeds this weak absorption by orders of magnitude. Since most of the experiments of photoluminescence excitation spectroscopy (PLE) in particular reflect this strong absorption, there is the risk that the experiments may not reveal the real extent of the band gap. A detailed and careful analysis of the band-gap of porous Si by means of PLE has been presented by Kux and Ben Chorin (1995). An average value of 0.2 eV above the luminescence peak energy has been found.

To summarize, the calculations result in corresponding inverse relationships between the energy of the band gap and the size of the nanoparticles and thus indicate quantum confinement. Correspondingly, in the optical spectra this inverse dependence can also be discovered which provides strong evidence for quantum confinement. In the next section some more experimental arguments will be given that support this hypothesis.

9.2 Silicon Nanocrystals and Quantum Structures in Porous Silicon

In search of convincing arguments for or against the quantum confinement hypothesis, one is confronted with several problems: (i) to prepare the nanostructures well enough to clearly identify the sizes, shapes and surface configurations, (ii) to assign the luminescence to the various channels of recombination and to prove its correlation to size variations, or (iii) to fit an *in*direct-gap absorption spectrum with suitable theories.

In what follows, we consider nanocrystals made from silicon (nc-Si) or occurring in layers of porous silicon (por-Si). Compared with the previously described synthesis of II–V compounds with ionic bonds and stabilization in coordinating solvents, the silicon structure is based on a strong covalently bonded diamond lattice. Therefore, the preparation is rather different from the one detailed in Chap. 2.

9.2.1 Preparation Methods

The most widely used preparation methods of nc-Si include radio frequency (rf) sputtering (Furukawa and Miyasato 1988; Yamamoto et al. 1991; Hayashi et al. 1989), laser evaporation (Hayashi et. al. 1990, 1993), and decomposition of silane gas by lasers, microwaves or high temperatures (Takagi et al. 1990; Kanemitsu et al. 1993a). The silicon nanocrystals thus obtained can be either free standing films or dispersed in colloids or polymers (Brus 1994; Brus et al. 1995; Gaponenko et al. 1994b; Bondarenko et al. 1994). The preparation of nc-Si from a thermal decomposition of disilane in high pressure He at temperatures close to 1000 °C has been reported by Littau et al. (1993).

The rf-sputtering method uses small pieces of silicon wafers, which are placed on a SiO_2 target and sputtered in an inert gas atmosphere. The size of the nanocrystals developing in the SiO_2 matrix can be controlled by the substrate temperature, the rf power and the gas pressure. A subsequent annealing at temperatures between 500 and 1000 °C is usually performed afterwards. Samples prepared by the rf-sputtering method contain silicon clusters with sizes around $\sim 2\,nm$, which can become larger during the annealing procedure (Hayashi et al. 1989). The luminescence spectrum of the as-deposited sample before annealing shows a peak at 1.75 eV (708 nm), which is shifted to lower energies depending on the temperature applied after the annealing. This shift of the luminescence peak correlated with the growth of the nc-Si particles has been explained as a size-dependent change of the energy gap (Furukawa and Miyasato 1988; Hayashi et al. 1989).[1]

The fabrication and size control of nc-Si by decomposition of silane gas (SiH_4) exploits the gas pressure dependence of the particle growth. Most of the silicon nanocrystals were collected on quartz plates for further characterization. The detection of oxidized silicon nanospheres produced by laser decomposition of silane has been reported by Kanemitsu et al. (1993a). The nc-Si spheres with radii between 3.5 and 7 nm show a broad luminescence band with the maximum at $\sim 1.65\,eV$ ($\sim 750\,nm$).

The pyrolysis at high temperatures close to 1000 °C followed by the dispersion of the crystallites into a colloid as proposed by Littau et al. (1993), is aimed at the growth of very small nc-Si spheres resulting from fast homogeneous nucleation in the hot aerosol. Stabilization is achieved by surface oxydation. The synthesis is performed in a three-zone reactor. In the first chamber of the reactor the pyrolysis of disilane occurs. The particle growth starts immediately after decomposition and nucleation by agglomeration of the formed clusters. The size can be controlled within limits via the initial gas concentration, the pressure, the temperature, and the time of the agglomeration process. The growth is stopped in the second chamber owing to continuous gas dilution. Reheating in the third part of the reactor causes the surface oxydation for stabilization. Finally the nanocrystals are collected

---

[1] For bulk crystalline silicon the band gap energy is 1.14 eV (1087 nm) at 300 K.

in ethylene glycol for further characterization. The nc-Si successfully isolated consist of a Si core with radii between 1.5 and 4 nm and a SiO_2 shell of constant thickness around 1 nm. Samples prepared according to that method have strong, broad emission with maxima around 800 nm, 725 nm and 660 nm for core radii of 1.55 nm, 0.85 nm, and 0.6 nm, respectively (Schuppler et al. 1994).

The analysis of the structures obtained by laser evaporation, silane decomposition or high temperature disilane pyrolysis revealed a capping of the surface with a layer of SiO_2 for the nc-Si spheres studied. Obviously, this surface layer strongly influences the optical properties of nc-Si. In the work of Kanemitsu et al. (1993a,b), the luminescence of the nc-Si is in fact shifted to higher energy, but for all samples by the constant value of 1.65 eV, although the dimensions of the crystalline Si cores were different. To explain this behavior, a model has been proposed where the electron–hole pairs are confined in the interface layer between the silicon core ($E_G = 1.14$ eV) and the SiO_2 shell ($E_G > 8$ eV). This structure may also give rise to quantum confinement, however the expected energy shifts are different from the simple quantum dot model. It resembles more closely the QDQW proposal presented in Chap. 2 and has been theoretically treated by Haus et al. (1993).

The fact that the structures which cause the quantum confinement can be very complicated is more clearly illustrated when we investigate porous silicon. Por-Si is usually obtained by electrochemical anodic oxidation of crystalline silicon (for a review see Smith and Kollins 1992, Lockwood 1994, Feng and Tsu 1994). The back of a silicon wafer is heavily doped and used as the anode of an electrolytic cell. The catode is provided by a metal (Pt) and the electrolyte for the anodization consists of a concentrated HF-solution. The anodic treatment with current densities of some tens of mA/cm^2 over some minutes produces a porous layer characterized by the size distribution of pores and the overall porosity. At very high values of porosity (≈ 70–80%), very small structures can be prepared with lateral dimensions sufficiently small (2–5 nm) to expect quantum confinement. The thickness of the porous layer is mainly controlled by the duration of the etching procedure, whereas the porosity is determined by the current densities. The surface of the pores is either saturated with hydrogen atoms from the HF solution or capped by an oxide shell. The manufacturing process and the width of the remaining silicon pillars implies on first sight the creation of an array of quantum wires, as suggested by Canham (1990), Lehmann and Gösele (1991), Bsiesy et al. (1991), and Xie et al. (1992).

9.2.2 Luminescence and Luminescence Dynamics

All silicon nanostructures of nanometer sizes which were prepared by the methods mentioned above show photoluminescence in the visible range. The high-energy shift of the luminescence reported in the literature for a great variety of nc-Si from different sources indicates quantum confinement, but the

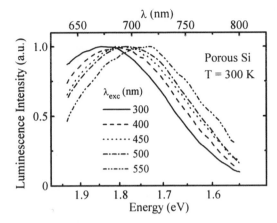

Fig. 9.2. Luminescence of porous Si excited at different energies. From Bondarenko et al. (1994)

detailed nanostructure still remains open. The results presented, for example, by Schuppler et al. (1994) and Hayashi et al. (1989) give evidence that the luminescence from the core exhibits quantum confinement, but the nanocrystals which emit in the visible range of the spectra (< 700 nm) are rather small particles.

Back in 1984 Pickering et al. reported strong visible, low-temperature photoluminescence of por-Si with efficiencies around 10 %. The emission of por-Si typically appears in the range between 600 and 800 nm (1.5–2 eV) depending on fabrication, storage and excitation conditions. An example of luminescence spectra obtained for por-Si with a porosity of the sample $\rho_0 = 30 \%$ is given in Fig. 9.2 (Bondarenko et al. 1994). The half-width of the band is about 0.3 eV at room temperature and does not show a noticeable narrowing at lower temperatures. The variation of the excitation energy results in a continuous shift of the emission spectrum and demonstrates the strong inhomogeneous character of the band.

Since the discovery of the intense red luminescence in por-Si, its microscopic origin has been a matter of controversy. Besides the quantum confinement models (Canham 1990; Lehmann and Gösele 1991; Xie et al. 1992; Vial et al. (1992); Koch et al. 1993; Lockwood and Wang 1995), other mechanisms providing visible luminescence in por-Si have been proposed such as the formation of siloxene (Stutzmann et al. 1992, 1993) or of amorphous silicon and SiH_x species (Vasques et al. 1992; Prokes 1993; Bondarenko et al. 1994; Gaponenko et al. 1994a). Thus, the microscopic nature of the visible emission is still the subject of intensive discussion, although several hundred papers have been published in this field since 1990 (for a review see e.g Lockwood 1994). Interestingly, all kinds of these structures actually seem to exist in porous Si layers.

Siloxene ($Si_6O_3H_6$) is a crystalline chemical compound based on Si_6 rings. It exists in different modifications depending on the positions of the oxygen atoms and likewise forms derivatives by substituting the hydrogen ligands by OH and other groups. Siloxene can be grown epitaxially on Si substrates, as demonstrated by Stutzmann et al. (1993). Because these thin layers of siloxene show strong green emission, the authors have ascribed the visible luminescence in por-Si to siloxene derivatives. In fact, depending on the preparation procedure, it could be possible that por-Si contains a fraction of siloxene in different modifications. The same authors carried out a detailed comparison of luminescence and PLE spectra between samples of siloxene and por-Si from different sources. Indeed, after an annealing procedure at 400 °C, siloxene shows spectra similar to those of por-Si. However, the high temperature might modify the siloxene, i.e., oxydation and transformation to SiO_2-covered Si clusters cannot be ruled out.

In the literature (see for example Lockwood 1994, Schuppler et al. 1994) there are various arguments against siloxene as the origin of the red luminescence, such as the size dependence, the missing signal from Si–O bonds in X-ray absorption or the instability of siloxene and its derivatives at high temperatures. The contribution from different types of amorphous silicon, however, cannot be completely ruled out. The PLE spectrum of por-Si samples investigated by Gaponenko et al. (1994a) very much resembles the dielectric function of amorphous Si. Therefore, preference for hydrogenated α−Si:H or oxydized α−Si:O has been given to explain in part the visible emission in por-Si (Prokes 1993; Prokes et al. 1994; Gaponenko et al. 1994a).

Over the last few years, a growing consensus has emerged that at least parts of the red emission are caused by quantum confinement. A convincing proof has been given by Schuppler et al. (1994) by comparing the luminescence of por-Si with different degrees of porosity and that of nc-Si with different sizes of the crystalline core. A size-dependent high-energy shift of the luminescence maxima has been found in both the por-Si and nc-Si samples. The spectra of both species show a close correlation if the sizes are similar, independent of how the structures were made. Quantum confinement by dot-like structures has been favored over extended wire arrays. At the present time, the exact structure of the network of pores is still not completely understood and the formed surface has not yet been identified. The important role of the surface is certainly an inherent characteristic of the structures. Both peak energy and efficiency of the luminescence are very sensitive to surface chemistry, particularly to the passivation by hydrogen or oxygen (Kovalev et al. 1994; Tsybeskov and Fauchet 1994a; Kanemitsu et al. 1993b). A very realistic model explaining the optical properties of nanostructures made from the indirect-gap semiconductor Si could be a quantum dot model combined with localization effects of the electrons or holes in the interface layer. The red emission starts from quantum confined states after relaxation of the optical excitation whereas the PLE reflects the quantum confined systems before relaxation. However, further research is necessary to tell a pure vibronic Stokes shift from a shift related to localization effects by states created by different surface chemistry.

The luminescence decay of por-Si has been studied on time scales ranging from picoseconds to milliseconds (Gardelis et al. 1991; Vial et al. 1992; Matsumoto et al. 1993; Suemoto et al. 1994; Gaponenko et al. 1994a). The decay time decreases with increasing temperatures and depends on the emission wavelength, being longer at longer wavelengths. The decay curves are nonexponential and mostly dominated by millisecond decay times.

Obviously, the straightforward idea that smaller sizes imply the more direct-gap character and, thus, fast and efficient emission is too simple. The value of the quantum yield in the red spectral region is more likely the result of the interplay between the enhancement of the radiative decay accompanied by different nonradiative mechanisms and could arise even from different species, for example, at the surface or in the interior of the nanostructure (Kovalev et al. 1994; Tsybeskov and Fauchet 1994a; Kanemitsu et al. 1993b, 1994; Matsumoto et al. 1993). Tunneling and escape of charge carriers can enhance fast nonradiative contributions (Vial et al. 1992). The evaluation of the transition probability in the theoretical work of Delerue et al. (1993) has led to the proposal that the high luminescence efficiency is due to the small probability of finding a nonradiative recombination center. In Calcott et al. (1993b) and Nash et al. (1995), a long living component observed particulary at low temperatures has been explained by the formation of singlet and triplet excitons in quantum-confined structures. This effect is based on the enhancement of the exchange splitting owing to the quantum confinement. The size dependence of the electron–hole exchange interaction in Si-nanocrystals has been theoretically treated for extremely small clusters by Takagahara and Takeda (1996) and a strong enhancement has been obtained.

A very successful treatment of the complex decay behavior of the luminescence has been presented by Suemoto et al. (1993, 1994). The authors start from the assumption of an inhomogeneous distribution in the radiative lifetimes. In this case, the nonexponential decay arises from the superposition of exponentials with different time constants. The tail of the measured decay curve then reflects the longest lifetime. This time constant has been attributed to the pure radiative lifetime and investigated in a first step. Next, fast nonradiative decay has been involved in the model to explain the nonexponential behavior at the beginning of the experimental decay curve at very early times. The nonradiative decay provides the possibility to explain the luminescence quenching at higher temperatures. To describe the nonradiative process, a sum of a tunneling process and a thermally activated escape of the carriers through the barrier has been assumed. The quantum confinement is reflected in different (Gauss distributed) barrier heights in the model.

Also in terms of micellar-like kinetics proposed by Gaponenko et al. (1994a), the spectral shift of the emission wavelength to lower energies can

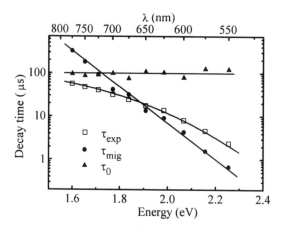

Fig. 9.3. Local radiative lifetime τ_0 (triangles) obtained from a modified stretched exponential model proposed by Gaponenko et al. (1996). The squares correspond to the measured red luminescence decay time τ_{exp} and the circles to the average migration time τ_{mig}, derived from the model

well be understood. For the fast nonradiative channel, a growing efficiency with decreasing size has been achieved simply by supposing the increase of the concentration of the trap centers, introduced for example, by an increase of the surface.

Wide spread fitting routines are stretched exponential functions like

$$I(t) = I_0 \exp -(t/\tau)^\beta \qquad (9.1)$$

to explain the luminescence dynamics in por-Si (for example Chen et al. 1992; Pavesi and Ceschini 1993; Ookubo 1993). In the frame of this model, the authors describe processes which are controlled by random walks of excitations in space with the dimensionality $d = 2\beta$. One example of such an analysis based on a modified stretched exponential function is shown in Fig. 9.3 (Gaponenko et al. 1996). Here the system is modeled by radiative centers walking randomly in the space with dimensionality d which undergo quenching during their migration. The mean decay time observed in the experiments is composed from the paramater τ_0 reflecting a local lifetime and the parameter τ_{mig}, which describes the deexcitation during the random walk in the space formed by the Si clusters. In this model, the local lifetime does not change, even if the *measured* lifetime decreases with increasing energy. Since the local lifetime represents the sum over all radiative and nonradiative processes it can be considered a lower limit for the radiative transition probability. Hence, a shortening of the luminescence decay time with decreasing size of the clusters is not necessarily a sign for an enhancement of radiative transition probability. For the dimensionality the value $d = 0.86$ has been derived, which is close to the dimensionality of a quantum wire.

9.2.3 Polarization of Luminescence

The importance of the surface configuration in nanostructures of silicon has been repeatedly emphasized in the course of this chapter. A further peculia-

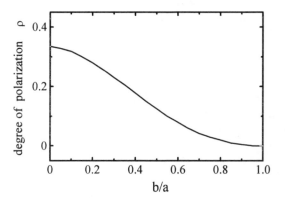

Fig. 9.4. Calculated degree of polarization of luminescence ρ as a function of the excentricity b/a of the ellipsoid. From Lavallard and Suris (1995)

rity introduced simply by the existence of a surface is the strong difference in the dielectric constants between the silicon particle and the environment. According to the discussion in Chap. 4, for high ratios ϵ_2/ϵ_1 and high volume fractions of semiconductor material $p \geq 1\%$, the local electric fields start to affect the optical properties. Because the strength of these local fields is strongly influenced by the geometry of the nanostructures, we get access to the shape of the emitting and absorbing entities by analysing the change in the polarization characteristics.

The luminescence of an isotropic, spherically symmetric system is unpolarized, because the absorbing and emitting dipoles are not correlated in space. A polarization of luminescence only occurs for resonantly excited optical transitions with dipole matrix elements sensitive to the polarization of the exciting light (selective excitation of states with certain symmetry). A polarization of luminescence at nonresonant excitation and after complete relaxation cannot be explained within such a model. In this case, the transition dipole must have an orientation in space due to local depolarization fields and the emission contains information about the shape of the nanostructure.

For wavelengths of light larger than the dimensions of the nanostructures the local fields can be described by electrostatics. For $\epsilon_2 > \epsilon_1$, a depolarization field appears in the vicinity of the nanostructure and reduces the amplitude inside the semiconductor material. Assuming an ellipsoidal or cylindrical shape of the nanostructures, the reduction depends on the angle between the incident light field and the main axis of the structure. The depolarization factor is smallest for light fields in the direction of the longest axis. The light, emitted by this ellipsoidal structure, is preferentially polarized in direction of that main axis.

Experimentally, the polarization of luminescence in por-Si has been discovered among others by Andrianov et al. (1993), Kovalev et al. (1995) and Gaponenko et al. (1995). The degree of linear polarization amounts to 20 % and is therefore a strong hint at structural deviations from spherical shape.

A model which correlates the high degree of polarization of luminescence to Si ellipsoids or Si chains has been proposed by Lavallard and Suris (1995). The authors calculated the degree of linear polarization for a random assembly of ellipsoids made from silicon. Figure 9.4 shows the degree of linear polarization $\rho = (I_z - I_x)/(I_z + I_x)$ of photoluminescence as a function of excentricity of the ellipsoid b/a for the ratio of dielectric constants $\epsilon_2/\epsilon_1 = 8.5$ ($\epsilon_2 = 15$ has been taken for the dielectric constant of silicon, and $\epsilon_1 = 1.77$ is the averaged dielectric constant of a porous Si layer with porosity of 74 %). I_z and I_x correspond to the intensities of luminescence polarized along z and x, whereas the exciting light has its polarization along z. The calculations were made for a rotational ellipsoid with the main axis a and equal values for the axes b and c. To obtain a degree of polarization around 20 %, the excentricity of the ellipsoid must be $b/a \sim 0.3$. Hence, the high degree of linear polarization can only be explained by the presence of wire-like structures in por-Si. A similar model has also been used by Ils et al. (1995) to explain the polarization of quantum wires etched from quantum wells in III–V materials. Moreover, the analysis of the degree of linear polarization provides the information about the orientation of the ellipsoids as shown by Kovalev et al. (1995). The observed anisotropy in the polarization properties of the por-Si samples investigated there has been ascribed to an aggregate of flattened and elongated ellipsoids. The elongated ellipsoids have their main axis in [100] direction and the flattened correspondingly perpendicular to that direction.

However, the explanation for the structures in por-Si of a complete quantum wire-model is very unlikely because the experimentally observed high-energy shift of the luminescence needs wire structures with dimensions below 1 nm diameter according to theoretical calculations (Hill and Whaley 1995b; Filonov et al. 1995). Such small structures already meet the limit of stability.

To summarize, the quantum confinement hypothesis can be considered as a well-established model, even if the dimension of the particles is smaller than expected (below 2 nm diameter) and hence more similar to clusters (Huaxiang 1993; Xie et al. 1994; Schuppler et al. 1994; Hayashi et al. 1989). In any case, the structure is determined by its surface configuration, the relative dimensions between shell and core and the material of the shell. The most unambiguous species that corresponds to a quantum dot seems to be the Si core surrounded by a closed shell of SiO_2. It would correspond to an indirect-gap semiconductor embedded in glass, but theoretically described by tight-binding models.

Nanocrystals made from Si and porous Si show very similar behavior in the spectra and dynamics of luminescence, but they differ distinctly in their polarization properties. For por-Si a strong linear polarization of luminescence has been reported, whereas polarization in luminescence is absent in spherical nc-Si. The most favored model for explaining the structure of porous Silicon is a network of various flattened or elongated nanocrystals, such as a mixture of dot-like and wire-like structures.

The luminescence decay does not reflect the strong change by orders of magnitudes of the oscillator strength as predicted by theory. The observed decay is strongly non-exponential with time constants between milliseconds and microseconds. The efficiency of emission, however, is substantial and reaches values around 10 % at room temperature and 50 % at low temperatures. The reason is given by the weaker nonradiative recombination, which is suppressed in Si nanostructures compared to the bulk silicon. In the spectra of emission, structures could also be revealed which were similar to the phonon-assisted transitions in the corresponding bulk indirect-gap material (Suemoto et al. 1993, 1994; Calcott et al. 1993b; Rosenbauer et al. 1995). These phonon replica, together with the relatively slow decay of luminescence, indicate that, in spite of the confinement effect, the silicon nanostructures maintain the indirect band-gap characteristics.

10. Concepts of Applications

The discovery of nonlinear and electro-optic phenomena in semiconductor-doped glasses and polymers very rapidly became associated with a growing technological interest. Regardless of the rather fragmentary understanding of 3D confinement, in the second half of the 1980s the proposals for applications were very optimistic and expectations were high concerning the potential of semiconductor-doped glasses for use in all-optical devices. At the beginning of the 1990s this opinion has been replaced by a more realistic evaluation of the practical use of quantum dots. Intensive efforts have been directed at the understanding and development of the technological processes. Since theory was still on a very high level, demand grew for better samples that allow comparison with the obtained theoretical results. During the last two to four years, the detailed concepts of the theory could be compared with experiments as a result of the further technological progress in sample preparation. A surprisingly high number of theoretical predictions could be verified in experiment and satisfactory correspondence exists, for example, for the calculations of the eigenenergies of one- and two-electron–hole-pair states for sizes of quantum dots around the Bohr radius.

Discussing concepts of applications in the following we restrict ourselves to three examples which might be promising. These are the use of quantum dots for

- fast optical switching, phase conjugation and wave guiding
- highly efficient emission of radiation and lasing with tunability over a wide spectral range
- microelectrodes and photocatalysts in photochemical reactions.

Furthermore, it should be mentioned that the single-electron transport expected from the lateral confinement of quantum dots by electric fields (see Chap. 7.2) might be a promising basis for binary logic operation; however, here we want to consider only properties related to optical applications. The optical properties considered in the following are thought to arise from the embedded semiconductor itself and not from the matrix material. However, the structure and composition of the matrix can control the material properties via the interface characteristics.

Let us turn to the first point. As demonstrated in Chap. 3, the energy states of the quantum dot can be deduced from the experiment in a very exact manner if analyzing the nonlinear absorption and its spectral behavior. When we exploit this nonlinear change of absorption or refraction, an optical Kerr shutter or Q-switch may be realized. At present, an intensive discussion persists about the practical implementation of semiconductor quantum dots, for example, as passive and active elements in integrated optics. Moreover, the development of new classes of nonlinear-optical materials has also led to the revival of concepts of optical computing. Recent investigations comprise, on the one hand, the search for 'figures of merit' for the different materials to facilitate the best choice from a multitude of suitable substances and, on the other hand, the development of entire device concepts favoring new materials, which are promising in the permanent competition with the advances in silicon technology. Special interest is dedicated to the role of quantum confinement and its influence on the nonlinear optical properties which could bring new application feasibilities.

The operation of the commercial filter glass Corning CS 3-69 as optical AND-gate has been reported by Acioli et al. (1990b). The on–off ratio of 1:10 was found for the transmission connected with switching times of 3 ps and an input fluence of $0.12 \, \mu J/cm^2$. Similar nonlinear changes in the refraction or absorption index have been reported by Yao et al. (1985) and Olbright and Peyghambarian (1986). The next consistent step towards optical logic elements is the combining of an optical nonlinearity with a suitable feedback resulting in optical bistability. Putting the nonlinear optical medium (here the semiconductor-doped glass) into a Fabry–Perot cavity, two different transmission values can be obtained for the same input light-intensity. This bistable behavior develops due to the internal feedback between optical path length and intensity in the cavity. By changing the input power, the refraction (and absorption) index changes, and thus the finesse of the cavity and/or the phase shift for one round trip are modulated. Eventually, the transmission switches between the two states. In Fig. 10.1 the optical bistable operation of a Fabry–Perot cavity containing CdS_xSe_{1-x}- doped glass as the nonlinear optical medium is seen with switching times of 25 ps and a switching power of $200 \, kW/cm^2$ ($T = 300 \, K$) (Yumoto et al. 1987).

Whereas in bulk semiconductors the physical origin of nonlinear absorption and refraction is associated with the optically created high-density many-particle system causing screening, band-gap renormalization and band filling, the mechanism of the nonlinearity in quantum dots is of a basically different kind. With decreasing size, the optical nonlinearity is strongly influenced by quantum confinement. In the case of quantum dot sizes around the bulk excitonic Bohr radius, the nonlinear optical response is dominated by the saturation of discrete levels of the one-electron–hole-pair states and the induced absorption of the two-pair states. These zero-dimensional structures can therefore be considered as experimental realization of discrete-level systems in semiconductors.

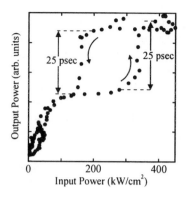

Fig. 10.1. Optical bistability using CdS_xSe_{1-x} -doped glass in a Fabry–Perot cavity ($T = 300\,\text{K}$, $\lambda_{\text{exc}} = 535\,\text{nm}$, $P_{\text{switch}} = 200\,\text{kW/cm}^2$). From Yumoto et al. (1987)

Owing to the different physical origin of the nonlinearity it is difficult to compare switching efficiency of the quantum dot and the bulk material. As far as the size of the dots is small and the energy states are well separated, the saturation intensity I_S (following the pattern of the two-level model) has often been used as the figure of merit to characterize the nonlinear switching behavior (see Chap. 6). Normally, we can expect that more than two levels are involved in the intensity-dependent absorption change, for example, as a result of interface-related trap states. Therefore the observed switching times are faster, but necessarily also connected to an increase of saturation intensity determining the switching power. In the present stage of technology, the switching energy necessary to achieve an efficient contrast is larger in CdS_xSe_{1-x}-doped glasses than those reported for the bulk materials (see for example Bohnert et al. 1984, Henneberger 1987, Broser and Gutowski 1988). The presently existing device concepts were mainly developed without special consideration of improvements owing to the quantum confinement. Here, the quantum size effect is mainly exploited for tuning the wavelength to one which is the best-suited for the switching operation.

Semiconductor-doped glasses have also been used in degenerate four-wave mixing (DFWM) experiments to demonstrate optical phase conjugation (Jain and Lind 1983; Roussignol et al. 1985, 1987a; Remillard et al. 1989). In the phase-conjugate geometry of the DFWM, two intensive counter-propagating pump beams and a weak probe beam (with small angle to the forward pump beam) are directed onto the sample. A fourth beam will be created by the induced polarizations and propagates in the direction opposite to the probe beam, with the field amplitude being the mathematical conjugate of the probe beam. Phase-conjugated mirrors can be used for aberration corrections of optical images. The potential of CdS_xSe_{1-x}-doped glasses for aberration corrections has been demonstrated by Jain and Lind (1983) and Remillard et al. (1989). Owing to the negligible diffusion processes in the doped-glasses, a phase-conjugated reflectivity of 0.3 % and additionally an increase of the field-of-view angle have been observed.[1]

The embedding of semiconductor quantum dots in soda-lime glasses opens a multitude of applications in waveguiding devices combining the passive and active elements in nonlinear optics. Switching in fibers, channel or planar waveguides has been the subject of several reviews (Jerominek et al. 1988; Ironside et al. 1988; Stegeman and Stolen 1989; Finlayson et al. 1988). Significant waveguide fabrication processes are the fiber drawing and the ion exchange in channel or planar waveguide configurations. These manufacturing steps provide optical beam confinement over long propagation distances and allow high-throughput intensities with moderate-power light sources. Semiconductor-doped fiber fabrication has been reported by Ainsli et al. (1987). A small rod of clear unannealed CdS_xSe_{1-x}-doped glass is inserted into a cladding sodium-calcium silica glass tube. The controlled formation of crystallites is achieved by local heat treatment in a 10 mm diameter central region of the 0.5 m long fiber. The fibers are few-moded and have typical losses of 5–7 dB/m at the edge wavelength of the filter (600 nm). The induced nonlinear refraction index shows in its time behavior a rapid recombination time of 20 ps followed by a long-time thermal component after applying an intensive pump pulse of 8 ps duration.

The fabrication of single mode planar and channel waveguides has been reported by Jerominek et al. (1988), Ironside et al. (1988), Stegeman and Stolen (1989), Finlayson et al. (1988), Cullen et al. (1986), and Gabel et al. (1987) using K^+/Na^+-ion exchange in soda-lime based glasses. The samples were heated in a bath of KNO_3 at 340 to 400 °C for 15 to 30 h (Cullen et al. 1986; Gabel et al. 1987). During the ion-exchange process the Na^+ ions in the glass were exchanged by the K^+ ions in the bath. The change of the refraction index near the surface obtained after this procedure is $\Delta n \approx 0.008$.

The coupling of light into the waveguide is realized by prism couplers or simply by end face coupling with efficiencies between 3 and 10 % and with low absorption losses of 0.5 dB/cm to 0.1 dB/cm. Two typical all-optical devices of interest based on waveguide fabrication are the directional coupler and the Mach–Zehnder interferometer (Stegeman and Stolen 1989; Finlayson 1988). Phase-conjugated reflection using ion-exchanged wave guides has been experimentally demonstrated by Gabel et al. (1987). Figure 10.2 shows the prism coupling of the two pump beams and the probe beam in the ion-exchanged waveguide of CdS_xSe_{1-x}-doped glass as well as the conjugate of the probe beam in DFWM configuration.

An optimized material for nonlinear waveguiding should exhibit large nonlinearity, fast response time and low absorption loss for the propagating light wave. To compare the II–VI doped glasses with other waveguide materials,

---

[1] The field of view is determined by the angle θ between the probe and the forward beam. An increase of the field of view means that also for large angles θ no washing-out and decrease of the signal intensity occurs.

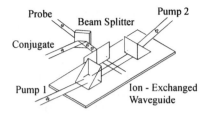

Probe

Conjugate

Pump 1

Beam Splitter

Pump 2

Ion - Exchanged
Waveguide

Fig. 10.2. Waveguide structure for optical phase conjugation using CdS_xSe_{1-x}-doped glasses. From Gabel et al. (1987)

a figure of merit has been proposed by Stegeman and Stolen (1989). Their parameter W

$$W = \Delta n_{\text{sat}} \frac{L}{\lambda} \frac{1}{L\alpha} = \frac{\Delta n_{\text{sat}}}{\lambda \alpha} \tag{10.1}$$

combines two requirements for optimum nonlinear waveguide device operation. The first term contains the necessary minimum phase shift of the refraction index Δn required for switching the waveguide device (L is the device length, λ is the wavelength). The second term is a so-called throughput criterion, ensuring that the damping losses due to absorption fulfill the requirement $L\alpha < 1$. When this figure of merit is applied to semiconductor-doped glasses, the nonlinear phase change observed is too small to make them good candidates for phase-dependent integrated optical switching, although the absorption losses are comparably low. For example, the value $W = 0.3$ obtained for CdS_xSe_{1-x}-doped glasses is confronted with $W > 10^3$ in multicomponent silica glasses containing highly polaric ions, or with organic materials ($W > 100$), where the lower nonlinearity is compensated for by extremely low absorption losses.

One possibility for bringing semiconductor-doped glasses into the realm of reality for use in nonlinear optical devices might be the investigation of *non*resonant optical nonlinearities. For light frequencies below the fundamental absorption edge in the so-called transparency region, the nonlinearity is associated with virtual excitations of bound electrons/holes exhibiting fast response times (model of the classical anharmonic oscillator). Cotter et al. (1992) argued that the specific effect of quantum confinement could enhance the nonresonant nonlinearity. The key is the theoretically predicted difference in the nonresonant two-photon absorption (TPA) coefficient for the quantum dot compared with the bulk semiconductor material. The authors propose that the probability of TPA is reduced in quantum dots and therefore the ratio $|\text{Re}\chi^{(3)}/\text{Im}\chi^{(3)}|$ can be enhanced with decreasing size, significantly altering the nonlinear-optical properties of the quantum-confined semiconductors if working in the transparency region. The predicted decreasing TPA absorption due to quantum confinement would open new possibilities for practical uses of devices based on the quantum-size effects.

The reported research work aimed at obtaining materials suitable for integrated optics is strongly encouraged by the vital interest in constructing optical computers. The advantage of optical methods does not lie solely in the computing velocity of the speed of light. The architecture of interconnects (photons do not interact if crossing), the increase of band width and the high number of spots communicating in parallel in optical imaging in principle offer more flexibility for computing than electronic circuits (for reviews see for example Smith 1984; Brenner 1988; Gibbs 1985). Experimental realizations of ideal two-level systems (i.e. binary systems) are therefore interesting for the rapidly developing field of optical computing and quantum cryptography.

A further important subject in the search for applications is the investigation of quantum yield and lasing in quantum dots. At first, the broad tuning range provided by the quantum dots is very attractive for light-emitting devices. In particular, silicon-based technology is searching for silicon structures that emit light with a higher quantum efficiency than bulk Si. In the context of this demand, porous Si has received most attention because of its visible emission with high quantum efficiency (1–10 %) at room temperature. The advantage of Si-based light-emitting devices is the high compatibility with the mature Si technology.

Light-emitting devices based on the diode concept demand materials providing both n- and p-type conductivity. The present development of chemical preparation methods for wide-gap semiconductor quantum dots makes the realization of blue-green light-emitting diodes combining the II–VI material with a conducting polymer (Wang and Herron 1992b) imaginable. For these materials, both electroluminescence as well as high quantum efficiency under optical pumping have already been demonstrated (Dabbousi et al. 1995, Colvin et al. 1994).

Another important effect is the observation of optical gain in quantum dots involving the potential for lasing devices. At first sight one would expect that the realization of an optical inversion could easily be achieved in the discrete level system of quantum dots. In the one-particle model the energy states are well-defined, the selection rules are clearly determined by $\Delta n = 0$ and $\Delta l = 0$ and the density of states is concentrated in a few levels. However, as already shown in the discussion of energy states in Chap. 3, this simple model of uncorrelated electron–hole pairs does not reflect the real situation in quantum dots. The Coulomb interaction and the band mixing change the selection rules and modify and split the energy states. One expects a very strong influence of the quantum-size effect on the laser characteristic, dependent on whether the strong or the weak confinement regime dominates (see Chap. 5). With decreasing size, major changes occur in interparticle interaction, transition rules, density of states, and influence of the competing interface capture.

One of the pioneering works in search of laser processes is the detection of gain by Dneprovskii et al. (1992b) in CdSe quantum dots. In samples of sizes around the excitonic Bohr radius, optical gain of $\sim 10\,\mathrm{cm}^{-1}$ at

$T = 80\,\mathrm{K}$ was observed and stimulated emission at the threshold energy of $0.5\,\mathrm{mJ/cm^2}$ was achieved for $\sim$30 ps pump pulses by inserting of the sample into a cavity of 1 mm in length. Stimulated emission has been reported for CuCl quantum dots in the weak confinement range of $R \approx 5\,\mathrm{nm}$ ($\approx 7a_B$) by Levy et al. (1991), Masumoto et al. (1993) and Faller et al. (1993). Gain in quantum dots with $R < a_B$ has been found by Giessen et al. (1995). A characteristic property of gain in strongly confined quantum dots is its spectral broadness of a few tens to hundreds of meV. Tremendous progress was achieved during the last year in developing lasing structures based on III–V materials. In particular, stimulated emission from buried quantum dots grown by epitaxial methods could be demonstrated at very low threshold current densities.

The last topic for discussion of application possibilities is related to II–VI quantum dots with organic colloids as the surrounding matrix material. Usually colloids or other organic environment provide a much better control of the interface region than glassy matrices. The proposed use of quantum dots in colloids for charge storage, microelectrodes or photocatalytic reactions contains very promising ideas which we are going to present in the following shortly.

Using a configuration where the organic matrix is laid on a conducting substrate and the transfer occurs through the quantum dot interface to that substrate, the quantum dot has been regarded as a small microelectrode (Henglein et al. 1988). The charged carriers created by the light led to an efficient photocurrent. The microelectrode properties have the prerequisite of high mobility of the charged carriers outside the dot and can therefore not be observed in an anorganic solid glassy matrix.

Ernsting et al. (1990) investigated fast electron ejection in quantum-sized $Zn_{1-x}Cd_xS$ colloidal solution using subpicosecond laser pump pulses. The yield of hydrated electrons was detected by their absorption at 616 nm, a typical wavelength in the spectrum of hydrated electrons in aqueous solution. The ejection of electrons was explained by an Auger-like process taking place after the population of the lowest electron–hole pair state. The time constant of 100 fs, found for this process is very fast.

An efficient electron ejection implies a remaining charged quantum dot, which should show a distinct shift in the absorption peak due to the electric field effect. This contradiction can be removed if it is taken into account, that chemical reactions are present in the semiconductor colloids. For example, holes can be oxydized by a free valence of S^{2-} ions or electrons can react with the cadmium ion finally causing a metal deposit at the interface.

In a sandwich structure the electron and holes are separated in space due to the tunneling of the electron photoexcited in semiconductor 1 into semiconductor 2 with the larger bandgap. The hole, however, remains in the semiconductor 1 material. This charge separation suggests long recombination times and possibly electro-optic response.

In conclusion, the progress in technology, and (in particular in the case of semiconductor-doped glasses) the wide distribution, availability, mechanical durability, and easy handling are serious reasons to continue paying attention to this class of materials. Benefits could be obtained from the charge separation associated with photorefractive, piezoelectric or other electro-optic effects, from the gradient in the refraction index exploiting wave-guide properties, from the realization of pn-junctions and from the suppression of nonradiative recombination enhancing the nonlinearity and the optical gain.

In spite of that, semiconductor-doped glasses can be considered very exciting materials for basic research. Although a quantum dot can be viewed as a large artificial atom, the distinct difference between a quantum dot and an atom consists in the much richer energy spectrum provided by the quantum dot as a result of the additional symmetry introduced by the Bloch part of the wavefunction.

References

Abraham, F.F. (1974): Homogeneous Nucleation Theory, Academic Press, New York.

Acioli, L.H., A.S.L. Gomes, J.M. Hickmann, and C.B. de Araujo (1990a): Appl. Phys. Lett. **56**, 2279.

Acioli, L.H., A.S.L. Gomes, J.R. Rios Leite, and C.B. De Araujo (1990b): IEEE J. Quantum Electr. **26**, 1277.

Ainsli, B.J., H.P. Girdlestone, D. Cotter (1987): Electr. Lett. **23**, 405.

Ando, K., Y. Yamada, and V.A. Shakin (1993): Phys. Rev. B **47**, 13462.

Andreani, L.C., and F. Bassani (1990): Phys. Rev. B **41**, 7536.

Andrianov, A.V., D.I. Kovalev, N.N. Zinovev, and I.D. Yaroshetskii (1993): JETP Lett. **58**, 427.

Alivisatos, A.P., T.D. Harris, P.J. Caroll, M.L. Steigerwald, and L.E. Brus (1989): J. Chem. Phys. **90**, 3463.

Alivisatos, A.P., T.D. Harris, N.J. Levinos, M.L. Steigerwald, and L.E. Brus (1988): J. Chem. Phys. **89**, 4001.

Artemyev, M.V., S.V. Gaponenko, I.N. Germanenko, and A.M. Kapitonov (1995): Chem. Phys. Lett. **243**, 450.

Ashcroft, N.W. and D.N. Mermin (1976): Solid State Physics, Saunders College Publishing, Int. Edition, Philadelphia 1976, p.270.

Bagnall, C.M. and J. Zarzycki (1990): J. Noncryst. Sol. **121**, 221.

Baldereschi, A. and N.O. Lipari (1971): Phys. Rev. B **3**, 439.

Baldereschi, A. and N.O. Lipari (1973): Phys. Rev. B **8**, 2697.

Banyai, L. (1989): Phys. Rev. B **39**, 8022.

Banyai, L., P. Gilliot, Y.Z. Hu, and S.W. Koch (1992): Phys. Rev. B **45**, 14136.

Banyai, L., Y.Z. Hu, M. Lindberg, and S.W. Koch (1988a): Phys. Rev. B **38**, 8142.

Banyai, L. and S.W. Koch (1993): Semiconductor Quantum Dots, Series on Atomic, Molecular and Optical Physics Vol. 2, World Scientific, Singapore.

Banyai, L., M. Lindberg, and S.W. Koch (1988b): Opt. Lett. **13**, 212.

Bassani, F., F. Tassone, and L.C. Andreani (1993): Proc. of the Internat. School Semiconductor Superlattices and Interfaces, ed. A. Stella and L. Miglio, North-Holland, p. 187.

Bastard, G. (1988): Wave Mechanics Applied to Semiconductor Heterostructures, Les Editions de Physique, Paris.

Bawendi, M.G., P.J. Caroll, W.L. Wilson, and L.E. Brus (1992): J. Chem. Phys. **96**, 946.

Bawendi, M.G., A.R. Kortan, M.L. Steigerwald, and L.E. Brus (1989): J. Chem. Phys. **91**, 7282.

Bawendi, M.G., M.L. Steigerwald, and L.E. Brus (1990a): Annual Rev. Phys. Chem. **41**, 477.

Bawendi, M.G., W.L. Wilson, L. Rothberg, P.J. Caroll, T.M. Jedju, M.L. Steiger-wald, and L.E. Brus (1990b): Phys. Rev. Lett. **65**, 1623.

Bayer, M., V.B. Timofeev, T. Gutbrod, A. Forchel, R. Steffen, and J. Oshinowo (1995): Phys. Rev. B **52**, R11623.

Beadie, G., E. Sauvain, A.S.L. Gomes, and N.M. Lawandy (1995): Phys. Rev. B **51**, 2180.

Becerra, L.R., C.B. Murray, R.G. Griffin, and M.G. Bawendi (1994): J. Chem .Phys. **100**, 3297.

Becker, R. and W. Döring (1935): Ann. Phys. (Leipzig) **24**, 719.

Beenakker, C.W.J. (1991): Physical Review B **44**, 1646.

Beenakker, C.W.J. and H. Van Houten (1991): Solid State Physics **44**, 1.

Belleguie, L. and L. Banyai (1991): Phys. Rev. B **44**, 8785.

Benisty, H. (1995): Phys. Rev. B **51**, 13281.

Benisty, H., C.M. Sotomayor-Torres, and C. Weisbuch (1991): Phys. Rev. B **44**, 10945.

Bhargava, R. N., D. Gallagher, X. Hong, and A. Nurmikko (1994): Phys. Rev. Lett. **72**, 416.

Bigot, J.Y., M.T. Portella, R.W. Schoenlein, C.V. Shank, and J.E. Cunningham (1990): In: C.B. Harris, E.P. Ippen, G.A. Mourou, and A. H. Zewail, 'Ultrafast Phenomena VII', Springer, Berlin, p. 239.

Binder, R., D. Scott, A.E. Paul, M. Lindberg, K. Henneberger, and S.W. Koch (1992): Phys. Rev. B, **45**, 1107.

Bloembergen, N. (1977): Nonlinear Optics, 3rd Ed. Benjamin, New York.

Bockelmann, U. (1993): Phys. Rev. B **48**, 17637.

Bockelmann, U. and G. Bastard (1990): Phys. Rev. B **42**, 8947.

Bockelmann, U. and T. Egeler (1992): Phys. Rev. B **46**, 15574.

Bohnert, K., F. Fidorra, and C. Klingshirn (1984): Z. Physik B **57**, 263.

Bondarenko, V.P., V.E. Borisenko, A.M. Dorofeev, I.N. Germanenko, and S.V. Gaponenko (1994): J. Appl. Phys. **75**, 2727.

Borelli, N.F., D. W. Hall, H.J. Holland, and D.W. Smith (1987): J. Appl. Phys. **61**, 5399.

Borelli, N.F. and D.W. Smith (1994): J. Noncryst. Solids **180**, 25.

Born M. and E. Wolf (1959): Principles of Optics, Pergamon Press, London.

Bottcher, C.J.F. (1973): Theory of Electric Polarization, 2nd ed. Vol. 1, Elsevier, Amsterdam.

Bowen Katari, J.E., V.L. Colvin, and A.P. Alivisatos (1994): J. Phys. Chem. **98**, 4109.

Brenner, K.H. (1988): Appl. Phys. B **46**, 111.

Broser, I. and J. Gutowski (1988): Appl. Phys. B **46**, 1.

Bruggeman, D.A.G. (1935): Ann. Phys. (Leipzig) **24**, 636.

Brunner, K., U. Bockelmann, G. Abstreiter, M. Walther, G. Böhn, G. Tränkle, and G. Weimann (1992): Phys. Rev. Lett. **69**, 3216.

Brus, L.E. (1984): J. Chem. Phys. **80**, 4403 (1984); **90**, 2555 (1986).

Brus, L.E. (1986): IEEE J. Quantum Electron. **22**, 1909.

Brus, L.E. (1991): Appl. Phys. A, **53**, 465.

Brus, L.E. (1994): J. Phys. Chem. **98**, 3575.

Brus, L.E., P.F. Szajewski, W.L. Wilson, T.D. Harris, S. Schuppler, and P.H. Citrin (1995): J. Am. Chem. Soc. **117**, 2917.

Bryant, G.W. (1993): Physica B **189**, 34.

Bsiesy, A., J.C. Vial, F. Gaspard, R. Herino, M. Lidgeon, F. Muller, R. Romestain, A. Vasiela, A. Halimaoui, and G. Bomchil (1991): Surf. Sci. **254**, 195.

Calcott, P.D.J., K.J. Nash, L.T. Canham, M.J. Kane, and D. Brumhead (1993a): J. Phys.: Condens. Matter **5**, L91.

Calcott, P.D.J., K.J. Nash, L.T. Canham, M.J. Kane, and D. Brumhead (1993b): J. Lumin. **57**, 257.

Campbell, I.H. and P.M. Fauchet (1986): Sol. State Comm. **58**, 739.

Canham, L.T. (1990): Appl. Phys. Lett. **57**, 1046.

Cardona, M. (1982): In: Light Scattering in Solids II, M. Cardona, and G. Güntherodt (eds.), Springer Berlin.

Cardona, M., K.L. Shaklee, and F.H. Pollak (1967): Phys. Rev. **154**, 696.

Chakraborty, T. (1992): Comments Cond. Mat. Phys. **16**, 35.

Chakraverty, B.K. (1967): J. Phys. Chem. Solids **28**, 2401 and 2413.

Chamarro, M.A., C. Gourdon, and P. Lavallard (1992): Sol. State Comm. **84**, 967.

Chamarro, M.A., C. Gourdon, P. Lavallard, and A.I. Ekimov (1995): Jpn. J. Appl. Phys. **34**, Suppl. 34-1, 12.

Chamarro, M.A., C. Gourdon, P. Lavallard, O. Lublinskaya, and A.I. Ekimov (1996): Phys. Rev. B, **53**, 1336.

Chamberlain, M.P., C. Trallero-Giner, and M.Cardona (1995): Phys. Rev. B **51**, 1680.

Champagnon, B., B. Andrianasolo, A. Ramos, M. Gandais, M. Allais, and J.-P. Benoit (1993): J. Appl. Phys. **73**, 2775.

Champion, P.M., G.M. Korenowski, and A.C. Albrecht (1979): Sol. State Comm. **32**, 7.

Chen, X., B. Henderson, and K.P. O'Donnell (1992): Appl. Phys. Lett. **60**, 2672.

Chen, Y., X.W. Lin, Z. Liliental-Weber, J. Wasburn, J.F. Klem, and J.Y. Tsao (1996): Appl. Phys. Lett. **68**, 111.

Chemla, D.S., and D.A.B. Miller (1986): Optics Lett. **11**, 522.

Chepic, D.I., Al.L. Efros, A.I. Ekimov, M.G. Ivanov, V.A. Kudriavtsev, and T.V. Yazeva (1990): J. Lum. **47**, 113.

Chestnoy, N., R. Hull, and L.E. Brus (1986): J. Chem. Phys. **85**, 2237.

Cho, K. (1976): Phys. Rev. B **14**, 4463.

Chiba, Y. and S. Onishi (1988): Phys. Rev. B **38**, 12988.

Colvin, V.L. and A.P. Alivisatos (1992a): J. Chem. Phys. **97**, 730.

Colvin, V.L., A.N. Goldstein, and A.P. Alivisatos (1992b): J. Amer. Chem. Soc. **114**, 5221.

Colvin, V.L., M.C. Schlamp, and A.P. Alivisatos (1994): Nature **370**, 354.

Cotter, D., M.G. Burt, and R.J. Manning (1992): Phys. Rev. Lett. **68**, 1200.

Cotter, D., H.P. Girdlestone, and K. Moulding (1991): Appl. Phys. Lett. **58**, 1455.

Cruz, de la, R.M., S.W. Teitsworth, and M.A. Stroscio (1995): Phys. Rev. B **52**, 1489.

Cullen, T.J., C.N. Ironside, C.T. Seaton, and G.I. Stegeman (1986): Appl. Phys. Lett. **49**, 1403.

Dabbousi, B.O., M.G. Bawendi, O. Onitsuka, and M.F. Rubner (1995): Appl. Phys. Lett. **66**, 1316.

D'Andrea, A., R. Del Sole, R. Girlanda, and A. Quattropani (eds.) (1992): Optics of Excitons in Confined Systems, IOP Conf. Ser. **123**, IOP Publishing Ltd. Bristol.

Daiminger, F., A. Schmidt, K. Pieger, F. Faller, and A. Forchel (1994): Semicond. Sci. Technol. **9**, 896.

Danielzik, B., K. Nattermann, and D. von der Linde (1985): Appl. Phys. B **38**, 31.

Davydov, A.S. (1987): Quantenmechanik, Akademieverlag Berlin.

Delerue, C., G. Alan, and M. Lannoo (1993): Phys. Rev. B **48**, 11024.

Demel, T., D. Heitmann, P. Grambow, and K. Ploog (1990): Phys. Rev. Lett. **64**, 788.

Demtröder, W. (1991): Laser Spectroscopy, Springer Berlin.

Dneprovskii, V.S., V.I. Klimov, D.K. Okorokov, and Yu.V. Vandyshev (1992a): phys. stat. sol. (b) **173**, 405.

Dneprovskii, V.S., V.I. Klimov, D.K. Okorokov, and Y.V. Vandyshev (1992b): Sol. State Commun. **81**, 227.

Dow, J.D. and D. Redfield (1970): Phys. Rev. B **1**, 3358.

Duval, E., A. Boukenter, and B. Champagnon (1986): Phys. Rev. Lett. **56**, 2052.

Edamatsu, K., S. Iwai, T. Itoh, S. Yano, and T. Goto (1995): Phys. Rev. B **51**, 11205.

Efros, Al.L. (1992a): Phys. Rev. B **46**, 7448.

Efros, Al.L. (1992b): Superlatt. Microstruct. **11**, 167.

Efros, Al.L. and A.L. Efros (1982): Sov. Phys. – Semicond. **16**, 772.

Efros, Al.L., A.I. Ekimov, F. Kozlowski, V. Petrova-Koch, H. Schmidbaur, and S. Shumilov (1991): Sol. State Comm. **78**, 853.

Efros, Al.L., V.A. Kharchenko, and M. Rosen (1995): Solid State Comm. **93**, 281.

Efros, Al.L., and A.V. Rodina (1989): Solid State Comm. **72**, 645.

Efros, Al.L., and A.V. Rodina (1993): Phys. Rev. B **47**, 10005.

Eichler, J.P., P. Günther, D.W. Pohl, (1986): Laser Induced Gratings, Springer Series in Optical Sciences **50**.

Einevoll, G.T. (1992): Phys. Rev. B **45**, 3410.

Ekimov, A.I. (1991): Physica Scripta **T39**, 217.

Ekimov, A.I., and Al.L. Efros (1988): phys. stat. sol. (b) **150**, 627.

Ekimov, A.I., Al.L. Efros, M.G. Ivanov, A.A. Onushenko, and S.K. Shumilov (1989): Sol. State Comm. **69**, 565.

Ekimov, A.I., Al.L. Efros, and A.A. Onushenko (1985a): Sol. State Comm. **56**, 921.

Ekimov, A.I., Al.L. Efros, T.V. Shubina, and A.P. Skvortsov (1990b): J. Lum. **46**, 97.

Ekimov, A.I., F. Hache, M.C. Schanne-Klein, D. Ricard, C. Flytzanis, I.A. Kudryavtsev, T.V. Yazeva, A.V. Rodina, Al. L. Efros (1993): J. Opt. Soc. Am. B **10**, 100.

Ekimov, A.I., I.A. Kudryavtsev, M.G. Ivanov, and Al. L. Efros (1990a): J. Lum. **46**, 83.

Ekimov, A.I. and A.A. Onushenko (1984): JETP Lett. **40**, 1137.

Ekimov, A.I., A.A. Onushenko, A.G. Plyukhin, and Al.L. Efros (1985b): Sov. Phys. JETP **61**, 891.

Ekimov, A.I., A.A. Onushenko, M.E. Raikh, and Al. L. Efros (1986): Sov. Phys. JETP **63**, 1054.

Ernsting, N.P., M. Kaschke, H. Weller, and L. Katsikas (1990): J. Opt. Soc. Am. B **7**, 1630.

Esch, V., B. Fluegel, G. Khitrova, H.M. Gibbs, Xu Jiajin, K. Kang, S.W. Koch, L.C. Liu, S.H. Risbud, and N. Peyghambarian (1990): Phys. Rev. B **42**, 7450.

Esch, V., K. Kang, B. Fluegel, Y.Z. Hu, G. Khitrova, H.M. Gibbs, S.W. Koch, N. Peyghambarian, L.C. Liu, and S.H. Risbud (1992): J. Nonlin. Opt. Phys. **1**, 25.

Fafard, S., R. Leon, D. Leonard, J.L. Merz, and P.M. Petroff (1994a): Phys. Rev. B **50**, 8086.

Fafard, S., D. Leonard, J.L. Merz, and P.M. Petroff (1994b): Appl. Phys. Lett. **65**, 1388.

Faller, P., B. Kippelen, B. Hönerlage, and R. Levy (1993): Opt. Mat. **2**, 39.

Farfad, S., R. Leon, D. Leonard, J.L. Merz, and P.M. Petroff (1995): Phys. Rev. B **52**, 5752.

Feng, Z.C. and R. Tsu (1994): Porous Silicon, World Scientific, Singapore.

Fick, J., G. Vitrant, A. Martucci, M. Guglielmi, S. Pelli, and G.C. Righini (1995): Nonlinear Optics **12**, 203.

Filonov, A.B., G.V. Petrov, V.A. Novikov, and V.E. Borisenko (1995): Appl. Phys. Lett. **67**, 1090.

Finlayson, N., W.C. Banyai, E.M. Wright, C.T. Seaton, G.I. Stegeman, T.J. Cullen, and C.N. Ironside (1988): Appl. Phys. Lett. **88**, 1144.

Fluegel, B., A. Paul, K. Meissner, R. Binder, S.W. Koch, N. Peyghambarian, F. Sasaki, T. Mishina, and Y. Masumoto (1992): Solid State Comm. **83**, 17.

Forchel, A. (1988): Advances in Solid State Physics (Festkörperprobleme) **28**, 99.

Franz, W. (1958): Z. Naturforsch. **13a**, 484.

Fröhlich, D., M. Haseloff, K. Reimann, and T. Itoh (1995): Sol. State Comm. **94**, 189.

Fujii, M., T. Nagareda, S. Hayashi, and K. Yamamoto (1991): Phys. Rev. B **44**, 6243.

Furukawa, S. and T. Miyasato, Phys. Rev. B **38**, 5726.

Gabel, A., K.W. DeLong, C.T. Seaton, and G.I. Stegeman (1987): Appl. Phys. Lett. **51**, 1682.

Gammon, D., E.S. Snow, B.V. Shanabrook, D.S. Katzer, and D. Park (1996): Phys. Rev. Lett. **76**, 3005.

Gaponenko, S.V., I.N. Germanenko, V.P. Gribkovskii, M.I. Vasil'ev, and V.A. Tsekhomskii (1992): SPIE Proc. **1807**, 65.

Gaponenko, S.V., I.N. Germanenko, E.P. Petrov, A.P. Stupak, V.P. Bondarenko, and A.M. Dorofeev (1994b): Appl. Phys. Lett. **64**, 85.

Gaponenko, S.V., V.K. Kononenko, E.P. Petrov, I.N. Germanenko, A.P. Stupak, and Y.H. Xie (1995): Appl. Phys. Lett. **67**, 3019.

Gaponenko, S.V., E.P. Petrov, U. Woggon, O. Wind, C. Klingshirn, Y.H. Xie, I.N. Germanenko, and A.P. Stupak (1996): J. Lum., to be published.

Gaponenko, S.V., U. Woggon, M. Saleh, W. Langbein, A. Uhrig, M. Müller, and C. Klingshirn (1993): J. Opt. Soc. Am. **10**, 1947.

Gaponenko, S.V., U. Woggon, A. Uhrig, W. Langbein, and C. Klingshirn (1994a): J. Lumin. **60**, 302.

Gardelis, S., J.S. Rimmer, P. Dawson, B. Hamilton, R.A. Kubiak, T.E. Whall, and E.H.C. Parker, (1991): Appl. Phys. Lett. **59**, 2188.

Gavazzi, G., J.M. Wheatley, and A.J. Schofield (1993): Phys. Rev. B **47**, 15170.

Geerligs, L.J., C.J.Harmans, and L.P.Kouwenhoven, (eds.) (1993): The Physics of Few-Electron Nanostructures, Physica B**189**, N1–4.

Gerthsen, D., K. Tillmann, and M. Lentzen (1994): Advances in Solid State Physics (Festkörperprobleme) **34**, 275.

Ghanassi, M., M.C. Schanne-Klein, F. Hache, A.I. Ekimov, D. Ricard, and C. Flytzanis (1992): Appl. Phys. Lett. **62**, 78.

Gibbs, H.M. (1985): Optical Bistability, Academic Press, New York.

Giessen, H., B. Fluegel, G. Mohs, Y. Z. Hu, N. Peyghambarian, and U. Woggon (1994): Int. Conf. Nonlinear Optics 94, Waikoloa, Hawaii, USA, Postdeadline-paper PD3; IEEE Nonlinear Optics (Institute of Electrical and Electronic Engineers, Piscataway, NJ, 1994) Vol. **1**, p. PD3B.

Giessen, H., B. Fluegel, G. Mohs, Y.Z. Hu, N. Peyghambarian, U. Woggon, C. Klingshirn, P. Thomas, and S.W. Koch (1996a): J. Opt. Soc. America B **13**, 1039.

Giessen, H., B. Fluegel, G. Mohs, N. Peyghambarian, J.R. Sprague, O.I. Micic, and A.J. Nozik (1996c): Appl. Phys. Lett. **68**, 304.

Giessen, H., N. Peyghambarian, and U. Woggon (1995): Optics and Photonics News **6**, No. 12. p.35; H. Giessen, U. Woggon , B. Fluegel, G. Mohs, Y.Z. Hu, S.W. Koch, and N. Peyghambarian (1996): Opt. Lett. 21, 1043.

Giessen, H., U. Woggon, B. Fluegel, G. Mohs, Y.Z. Hu, S.W. Koch, and N. Peyghambarian (1996b): J. Chem. Phys., in press.

Gilliot, P., J.C. Merle, R. Levy, M. Robino, and B. Hönerlage (1989): phys. stat. sol. (b) **153**, 403; P. Gilliot, PhD-Thesis, Universite Strasbourg 1990.

Gindele, F. (1996): Diploma theses University of Karlsruhe; Gindele, F., O. Wind, and U. Woggon (1996): Proc. Intern. Conf. Semicond. Phys. (ICPS 96), Berlin, to be published.

Goerigk, G., H.G. Haubold, C. Klingshirn, and A. Uhrig (1994): J. Appl. Cryst. **27**, 907.

Goetz, K.-H., D. Bimberg, H. Jürgensen, J. Selders, A.V. Solomonov, G.F. Glinskii, and M. Razeghi (1983): J. Appl. Phys. **54**, 4543.

Gogolin, O., Y. Berosashvili, G. Mschvelidze, E. Tsitsischvili, A. Uhrig, H. Giessen, and C. Klingshirn (1991): Sem. Sci. Technol. **6**, 401.

Golubkov, V.V., A.I. Ekimov, A.A. Onushenko, and V.A. Tsekhomskii (1981): Phys. Chem. Glass **7**, 397.

Gourgon, C., Le Si Dang, H. Mariette, C. Vieu, and F. Muller (1995): Appl. Phys. Lett. **66**, 1635.

Gutsche, E. and H. Lange (1967): phys. stat. sol.(b) **22**, 229.

Grigoryan, B., E.M. Kazaryan, Al.L. Efros and T.V. Yazeva (1990): Sov. Phys.–Solid State **32**, 1031.

Grun, J.B., C. Comte, R. Levy, and E. Ostertag (1976): J. Lum. **12/13**, 581.

Grundmann, M., J. Christen, N.N. Ledentsov, J. Böhrer, D. Bimberg, S.S. Ruvimov, P. Werner, U. Richter, U. Gösele, J. Heidenreich, V.M. Ustinov, A.Yu. Egorov, A.E. Zhukov, P.S. Kopev, and Zh. I. Alferov (1995a): Phys. Rev. Lett. **74**, 4043.

Grundmann, M., N.N. Ledentsov, O. Stier, D. Bimberg, V.M. Ustinov, P.S. Kopev, and Zh. I. Alferov (1996): Appl. Phys. Lett. **68**, 979.

Grundmann, M., U. Lienert, J. Christen, D. Bimberg, A. Fischer-Colbrie, and J.N. Miller (1990): J. Vac. Sci. Technol. **B 8**, 751.

Grundmann, M., O. Stier, and D. Bimberg (1995b): Phys. Rev. B **52**, 11969.

Hache, F., M.C. Klein, D. Ricard, and C. Flytzanis (1991): J. Opt. Soc. B **8**, 1802.

Hache, F., D. Ricard, and C. Flytzanis (1986): J. Opt. Soc. Am. B **3**, 1647.

Hache, F., D. Ricard, and C. Flytzanis (1989), Appl. Phys. Lett. **55**, 1504.

Hallam, L.D., J. Weis, and P.A. Maksym (1996): Phys. Rev. B **53**, 1452.

Halsall, M.P., J.E. Nicholls, J.J. Davies, B. Cockayne, and P.J. Wright (1992): J. Appl. Phys. **71**, 907.

Hase, N. and M. Onuki (1970): J. Phys. Soc. Japan **28**, 965.

Hatami, F., N.N. Ledentsov, M. Grundmann, J. Böhrer, F. Heinrichsdorff, M. Beer, D. Bimberg, S.S. Ruvimov, P. Werner, U. Gösele, J. Heydenreich, U. Richter, S.V. Ivanov, B. Ya. Meltser, P.S. Kopev, and Zh.I. Alferov (1995): Appl. Phys. Lett. **67**, 656.

Haug, H. and L. Banyai, (eds.) (1989): Optical Switching in Low-Dimensional Systems, Plenum Press, New York.

Haug, H. and S. W. Koch (1993): Quantum Theory of the Optical and Electronic Properties of Semiconductors, 2Ed., World Scientific Singapore.

Haug, R.J. (1994): Advances in Solid State Physics 34, 219.

Haus, J.W., H.S. Zhou, I. Honma, and H. Komiyama (1993): Phys. Rev. B **47**, 1359.

Hayashi, S., M. Fujii, and K. Yamamoto (1989): Jpn. J. Appl. Phys. **28**, L1464.

Hayashi, S. and H. Kanamori (1982): Phys. Rev. B **26**, 7079.

Hayashi, S., Y. Kanzawa, M. Kataoka, T. Nagareda, and K. Yamamoto (1993): Z. Phys. D **26**, 144.

Hayashi, S., S. Tanimoto, M. Fujii, and K. Yamamoto (1990): Superlattices Microstruct. **8**, 13.

Heitmann, D. (1995): Physica B **212**, 201.

Heitmann, D., and J.P. Kotthaus (1993): Physics Today **46**, 56 (6).

Heitz, R., M. Grundmann, N.N. Ledentsov, L. Eckey, M. Veit, D. Bimberg, V.M. Ustinov, A.Yu. Egorov, A.E. Zhukov, P.S. Kopev, and Zh. I. Alferov (1996): Appl. Phys. Lett. **68**, 361.

Hendershot, D.G., D.K. Gaskill, B.L. Justus, M. Fatemi, and A.D. Berry (1993): Appl. Phys. Lett. **63**, 3324.

Henglein, A. (1988): Top. Curr. Chemistry **143**, 115.

Henglein, A. (1989): Chem. Rev. **89**, 1861.

Henneberger, F. (1987): phys. stat. sol. (b) **137**, 371.

Henneberger, F., J. Puls, A. Schülzgen, V. Jungnickel, Ch. Spiegelberg (1992): Festkörperprobleme **32**, 279.

Henneberger, F., J. Puls, Ch. Spiegelberg, A. Schülzgen, H. Rossmann, V. Jungnickel, and A.I. Ekimov (1991): Semicond. Sci. Technol. **6**, A41.

Henneberger, F., U. Woggon, J. Puls, and Ch. Spiegelberg (1988): Appl. Phys. B **46**, 19.

Herron, N., J.C. Calabrese, W.E. Farneth, and Y. Wang (1993): Science **259**, 1426.

Herron, N., Y. Wang, and H. Eckert (1990): J. Amer. Chem. Soc. **112**, 1322.

Herron, N., Y. Wang, M.M. Eddy, G.D. Stucky, D.E. Cox, K. Moller, and T. Bein (1989): J. Am. Chem. Soc. **111**, 530.

Hilinski, E., P. Lucas, (1988): J. Chem. Phys. **89**, 3435.

Hill, N. and B. Whaley (1995a): Phys. Rev. Lett. **75**, 1130.

Hill, N. and B. Whaley (1995b): Appl. Phys. Lett. **67**, 1125.

Hoheisel, W., V.L. Colvin, C.S. Johnson, and A.P. Alivisatos (1994): J. Chem. Phys. **101**, 8455.

Hönerlage, B., R. Levy, J.B. Grun, C. Klingshirn, and K. Bohnert (1985): Physics Reports **124**, 161.

Hönerlage, B., U. Rössler, Vu Duy Phach, A. Bivas, and J.B. Grun (1980): Phys. Rev. B **22**, 797.

Hu, Y.Z., H. Giessen, N. Peyghambarian, and S.W. Koch (1996): Phys. Rev. B **53**, 4814.

Hu, Y.Z., S.W. Koch, M. Lindberg, N. Peyghambarian, E.L. Pollock, and F. Abraham (1990a): Phys. Rev. Lett. **64**, 1805.

Hu, Y.Z., M. Lindberg, and S.W. Koch (1990b): Phys. Rev. B **42**, 1713.

Huang, G.L., and H.S. Kwok (1992): J. Opt. Soc. Am. B **9**, 2019.

Huang, K., and A. Rhys (1950): Proc. Roy. Soc. London, Ser. A **204**, 406.

Huaxiang, F., Y. Ling, and X. Xide (1993): Phys. Rev. B **48**, 10978.

Hulin, D., A. Mysysrowicz, A. Migus, and A. Antonetti (1985): J. Lum. **30**, 290.

Hunsche, S., T. Dekorsy, V. Klimov, and H. Kurz (1996): Appl. Phys. B, to be published.

Hybertsen, M.S. (1994): Phys. Rev. Lett. **72**, 1514.

Illing, M., G. Bacher, T. Kümmell, A. Forchel, T.G. Andersson, D. Hommel, B. Jobst, and G. Landwehr (1995): Appl. Phys. Lett. **67**, 124.

Ils, P., Ch. Greus, A. Forchel, V.D. Kulakovskii, N.A. Gippius, and S.G. Tikhodeev (1995): Phys. Rev. B **51**, 4272.

Ils, P., M. Michel, A. Forchel, and I. Gyuro (1993): J. Vac. Sci. Technol. **11**, 2584.

Inoshita, T. and H. Sakaki (1992): Phys. Rev. B **46**, 7260.

Ironside, C.N., T.J. Cullen, B.S. Bhumbra, and J. Bell, W.C. Banyai, N. Finlayson, C.T. Seaton, and G.I. Stegeman (1988): J. Opt. Soc. Am. **5**, 492.

Itoh, T., Y. Iwabuchi, and M. Kataoka (1988a): phys. stat. sol. (b) **145**, 567.

Itoh, T., Y. Iwabuchi, and T. Kirihara (1988b): phys. stat. sol. (b) **146**, 531.

Itoh, T., F. Jin, Y. Iwabuchi, and T. Ikehara (1989): In: Nonlinear Optics of Organics and Semiconductors, ed. by T. Kobayashi, Springer, Berlin, p. 76.

Itoh, T., M. Nishijima, A.I. Ekimov, C. Gourdon, Al.L. Efros, and M. Rosen (1995): Phys. Rev. Lett. **74**, 1645.

Itoh, T., S. Yano, N. Katagiri, Y. Iwabuchi, C. Gourdon, and A.I. Ekimov (1994): J. Lum. **60/61**, 369.

Jain, R.K. and R.C. Lind (1983): J. Opt. Soc. Am. **73**, 647.

Jerominek, H., S. Patela, M. Pigeon, Z. Jakubczyk, C. Delisle, and R. Tremblay (1988): J. Opt. Soc. Am. **5**, 496.

Jorda, S., U. Rössler, and D. Broido (1992): Proceedings of the International Meeting on Optics of Excitons in Confined Systems, ed. A. D'Andrea, R. Del Sole, R. Girlanda, and A. Quattropani, The Institute of Physics Conf. Series **123**, 49.

Jungk, G. (1988a): phys. stat. sol. (b) **146**, 335.

Jungk, G. (1988b): phys. stat. sol. (b) **150**, 483.

Jungk, G. and U. Woggon (1991): Superlatt. Microstruct. **9**, 401.

Junker, P., U. Kops, U. Merkt, T. Darnhofer, and U. Rössler (1994): Phys. Rev. B **49**, 4794.

Jursenas, S., V. Stepankevicius, M. Strumskis, and A. Zukauskas (1995): Semic. Sci. Technol. **10**, 302.

Justus, B.L., R.J.Tonucci, and A.D.Berry (1992): Appl. Phys. Lett. **61**, 3151.

Kagan, C.R., C.B. Murray, M. Nirmal, and M.G. Bawendi (1996): Phys. Rev. Lett. **76**, 1517.

Kane, E.O. (1957): J. Chem. Phys. Solids **1**, 249.

Kanemitsu, Y., T. Futagi, T. Matsumotu, and H. Mimura (1994): Phys. Rev. B **49**, 14723.

Kanemitsu, Y., T.Ogawa, K.Shiraishi, and K.Takeda (1993a): Phys. Rev. B **48**, 4883.

Kanemitsu, Y., H. Uto, T. Matsumoto, T. Futagi, and H. Mimura (1993b): Phys. Rev. B **48**, 2827.

Kang, K.I., A.D. Kepner, S.V. Gaponenko, S.W. Koch, Y.Z. Hu, and N. Peyghambarian (1993): Phys. Rev. **48**, 15449.

Kang, K.I., A.D. Kepner, Y.Z. Hu, S.W. Koch, N. Peyghambarian, C.Y. Li, T. Takada, Y. Kao, and J.D. Mackenzie (1994): Appl. Phys. Lett. **64**, 1487.

Kang, K.I., B.P. McGinnes, Sandalphon, Y.Z. Hu, S.W. Koch, N. Peyghambarian, A. Mysyrowicz, L. C. Liu, and S.H. Risbud (1992): Phys. Rev. B **45**, 3465.

Kash, K. (1990): J. Lumin. **46**, 69.

Kastner, M. (1992): Rev. Mod. Phys. **64**, 849.

Katsikas, L., A.Eychmuller, M.Giersig, and H.Weller (1990): Chem. Phys. Lett. **172**, 201.

Kayanuma, Y. (1988): Phys. Rev. B **38**, 9797.

Kayanuma, Y. and H. Momiji (1990): Phys. Rev. B **41**, 10261.

Keldysh, L.V. (1958): Soviet phys. – JETP **7**, 788.

Khveshchenko, D.V., and S.V. Meshkov (1993): Phys. Rev. B **47**, 12051.

Kitamura, M., M. Nishioka, J. Oshinowo, and Y. Arakawa (1995): Appl. Phys. Lett. **66**, 3663.

Klein, M.C., F. Hache, D. Ricard, and C. Flytzanis (1990): Phys. Rev. B **42**, 11123.

Klimov, V., P. Haring Bolivar, and H. Kurz (1996): Phys. Rev. B **53**, 1463.

Klimov, V., S. Hunsche, and H. Kurz (1994): Phys. Rev. B **50**, 8110.

Klimov, V., S. Hunsche, and H. Kurz (1995): phys. stat. sol. (b) **188**, 259.

Klingshirn, C. (1995): Semiconductor Optics, Springer, Berlin.

Knipp, P.A., and T.L. Reinecke (1992): Phys. Rev. B **46**, 10310.

Koch, F., V. Petrova-Koch, and T. Muschik (1993): J. Lum. **57**, 271.

Koch, S.W. (1984): Dynamics of First Order Phase Transitions in Equlilibrium and Nonequlilibrium Systems, Lecture Note in Physics **207**, Springer, Berlin.

Koch, S.W., Y. Z. Hu, B. Fluegel, and N. Peyghambarian (1992): J. Crystal Growth **117**, 592.

Koda, T., Y. Segawa, and T. Mitani (1970): Phys. Status Solidi **38**, 821.

Kortan, A.R., R. Hull, R.L. Opila, M.G. Bawendi, M.L. Steigerwald, P.J. Caroll, and L.E. Brus (1990): J. Amer. Chem. Soc. **112**, 1327.

Kouwenhoven, L.P., N.C. van der Vaart, A.T. Johnson, C.J.P.M. Harmans, J.G. Williamson, A.A.M. Staring, and C.T. Foxon (1991): Advances in Solid State Physics **31**, 329.

Kovalev, D.I., M. Ben Chorin, J. Diener, F. Koch, Al.L. Efros, M. Rosen, N.A. Gippius, and S.G. Tikhodeev (1995): Appl. Phys. Lett. **67**, 1585.

Kovalev, D.I., I.D. Yaroshetzkii, T. Mushik, V. Petrova-Koch, and F. Koch (1994): Appl. Phys. Lett. **64**, 214.

Krost, A., F. Heinrichsdorff, D. Bimberg, A. Darhuber, and G. Bauer (1996): Appl. Phys. Lett. **68**, 785.

Kuchar, F., H. Heinrich, and G. Bauer (eds.) (1990): Localization and Confinement of Electrons in Semiconductors, Springer, Berlin.

Kulinkin, B.S., V.A.Petrovskii, V.A.Tsekhomskii, and M.F.Shchanov (1988): Phys. Chem. Glass **14**, 470.

Kux, A., and M. Ben Chorin (1995): Phys. Rev. B **51**, 17535.

Laheld, U.E.H., F.B. Pedersen, and P.C. Hemmer (1993): Phys. Rev. B **48**, 4659.

Laheld, U.E.H., F.B. Pedersen, and P.C. Hemmer (1995): Phys. Rev. B **52**, 2697.

Lamb, H. (1882): Proc. Math. Soc. London **13**, 187.

Landau, L.D. and E.M. Lifshitz (1974): Lehrbuch der Theoretischen Physik, Band VI, Akademie-Verlag, Berlin.

Landolt–Börnstein (1982): Physics of II-VI and I-VII Compounds, Vol. **17b**, New Series, Ed. O. Madelung, Springer-Verlag, Berlin/Heidelberg/New York.

Langbein, W., M. Hetterich, and and C. Klingshirn (1995): Phys. Rev. B **51**, 9922.

Langbein, W., D. Lüerssen, H. Kalt, W. Braun, and K. Ploog (1996): Phys. Rev. **53**, 15473; W. Langbein, PhD-thesis, University of Karlsruhe, Germany, 1995.

Lannoo, M., C. Delerue, and G. Allan (1995): Phys. Rev. Lett. **74**, 3415.

Lap-Tak Cheng, N. Herron, and Y. Wang (1989): J. Appl. Phys. **66**, 3417.

Lavallard, P. and R.A. Suris (1995): Solid State Comm. **95**, 267.

Lehmann, V. and U. Gösele (1991): Appl. Phys. Lett. **58**, 856.

Leon, R., S. Fafard, D. Leonard, J.L. Merz, and P.M. Petroff (1995): Appl. Phys. Lett. **67**, 521.

Leonard, D., M. Krishnamurthy, C.M. Reaves, S.P. DenBaars, and P.M. Petroff (1993): Appl. Phys. Lett. **63**, 3203.

Leonard, D., K. Pond, and P.M. Petroff (1994): Phys. Rev. B **50**, 11687.

Leonelli, R., A. Manar, J.B. Grun, and B. Hönerlage (1992): Phys. Rev. B **45**, 4141.

Letokhov, V.S. and V.P. Chebotaev (1977): Principles of Nonlinear Laser Spectroscopy, Springer Berlin.

Leung, K.M. (1986): Phys. Rev. A, **33**, 2461.

Levy, R., L. Mager, P. Gilliot, and B. Hönerlage (1991): Phys. Rev. B **44**, 11286.

Li, G.H., A.R. Goni, K. Syassen, O. Brandt, and K. Ploog (1994): Phys. Rev. B **50**, 18420.

Li, Y.Q., C.C. Sung, R. Inguva, and C.M. Bowden (1989): J. Opt. Soc. Am. B **6**, 814.

Lifshitz, I.M. and V.V. Slezov (1961): J. Phys. Chem. Solids **19**, 35.

Lippens, P.E. and M. Lannoo (1989): Phys. Rev. B **39**, 10935.

Lippens, P.E. and M. Lannoo (1990): Phys. Rev. B **41**, 6079.

Lipsanen, H., M. Sopanen, and J. Ahopelto (1995): Phys. Rev. B **51**, 13868.

Littau, K.A., P.J. Szajowski, A.J. Miller, A.R. Kortan, and L.E. Brus (1993): J. Phys. Chem. **97**, 1224.

Liu, Li-Chi and S.H. Risbud (1990): J. Appl. Phys. **68**, 28.

Lockwood, D.J. (1994): Sol. State Comm. **92**, 101.

Lockwood, D. and A. G. Wang (1995): Solid State Comm. **94**, 905.

Lubyshev, D.I., P.P. Gonzalez-Borrero, E. Marego, E. Petitprez, N. La Scala, and P. Basmaji (1996): Appl. Phys. Lett. **68**, 205.

Luong, J.C. and N.F. Borelli (1989): Mater. Res. Soc. Symp. Proc. **144**, 695.

Luttinger, J.M. (1956): Phys. Rev. **102**, 1030.

MacDonald, R.L., and N.M. Lawandy (1993): Phys. Rev. B **47**, 1961.

Madelung, O. (1972): Festkörpertheorie, Vol. 1, Springer, Berlin, Heidelberg, New York, p. 29 ff.

Maeda, Y. (1995): Phys. Rev. B **51**, 1658.

Malhotra, J., D.J. Hagan, and B.G. Potter (1991): J. Opt. Soc. Am. B **8**, 1531.

Marchetti, A.P., K.P.Johansson, and G.L.McLendon (1993): Phys. Rev. B **47**, 4268.

Marini, J.C., B. Stebe, and E. Kartheuser (1994): Phys. Rev. B **50**, 14302.

Marzin, J.Y. and G. Bastard (1994b): Sol. State Comm. **92**, 437.

Marzin, J.Y., J.M. Gerard, A. Israel, D. Barrier, and G. Bastard (1994a): Phys. Rev. Lett. **73**, 716.

Matsumoto, T., T. Futagi, H. Mimura, and Y. Kanemitsu (1993): Phys. Rev. B **47**, 13876.

Matsumoto, T., N. Hasegawa, T. Tamaki, K. Ueda, T. Futagi, H. Mimura, and Y. Kanemitsu (1994): Jap. J. Appl. Phys. **33**, L35.

Masumoto, Y., S. Katayanagi, and T. Mishina (1994a): Phys. Rev. B **49**, 10782.

Masumoto, Y., T. Kawamura, and K. Era (1993): Appl. Phys. Lett. **62**, 225.

Masumoto, Y., T. Kawamura, T.Ohzeki, and S.Urabe (1992): Phys.Rev. B **46**, 1827.

Masumoto, Y., S. Okamoto, and S. Katayanagi (1994b): Phys. Rev. B **50**, 18658.

Masumoto, Y., Y. Unuma, and S. Shionoya (1984): Semiconductors Probed By Ultrafast Laser Spectroscopy, Vol. 1, Academic Press, pp.307.

Masumoto, Y. T. Wamura, and A. Iwaki (1989a): Appl. Phys. Lett. **55**, 2535.

Masumoto, Y., M. Yamazaki, and H. Sugawara (1989b): Appl. Phys. Lett. **53**, 1527.

Masumoto, Y., L.G. Zimin, K. Naoe, S. Okamoto, T. Kawazoe, and T. Yamamoto (1995): J. Lum. **64**, 213.

Maxwell-Garnett (1906): J.C., Philos. Trans. R. Soc. London **203**, 385 (1904); Philos. Trans. R. Soc. London, Ser. A **205**, 237 (1906).

McEuen, P.L., N.S. Windgreen, E.B. Foxman, J. Kinaret, U. Meirav, M.A. Kastner, Y. Meir, and S.J. Wind (1993): Physica B **189**, 70.

Meissner, K., B. Fluegel, H. Giessen, B. P. McGinnis, R. Binder, A. Paul, S.W. Koch, and N. Peyghambarian (1993): Phys. Rev. B **48**, 15472.

Meissner, K., B. Fluegel, H. Giessen, G. Mohs, R. Binder, S.W. Koch, and N. Peyghambarian (1994): Phys. Rev. B **50**, 17647.

Merkt, U. (1990): Adv. Sol. State Physics **30**, 77.

Merkt, U. (1993): Physica B **189**, 165.

Merlin, R., G. Güntherodt, R. Humphreys, M. Cardona, R. Suryanarayanan, and F. Holtzberg (1978): Phys. Rev. B **17**, 4951.

Mews, A., A. Eychmüller, M. Giersig, D. Schooss, and H. Weller (1994): J. Phys. Chem. **98**, 934.

Meystre, P. and M. Sargent III (1990): Elements of Quantum Optics, Springer Berlin, Heidelberg, New York.

Miller, D.A.B., D.S. Chemla, T.C. Damen, T.C. Gossard, W. Wiegmann, T.H. Wood and C.A. Burrus (1984): Phys. Rev. Lett. **53**, 2173; Phys. Rev. B **32**, 1043 (1985).

Miller, D.A.B., D.S. Chemla, T.C. Damen, T.H. Wood, C.A. Burrus, A.C. Gossard, and W. Wiegmann (1985): IEEE J. Quantum Electr. **QE 21**, 1462.

Miller, D.A.B., D.S. Chemla, and S. Schmitt-Rink (1988a): Optical Nonlinearities and Instabilities in Semiconductors, Ed. H. Haug, Academic Press, pp. 325.

Miller, D.A.B., D.S. Chemla, and S. Schmitt-Rink (1988b): Appl. Phys. Lett. **52**, 2154.

Misawa, K., T. Hattori, T. Kobayashi, Y. Ohashi, and H. Itoh (1988): Nonlinear Optics of Organics and Semiconductors, ed. T. Kobayashi, Springer, Berlin, p. 66.

Misawa, K., S. Nomura, and T. Kobayashi (1992): Optical Properties of Nano-structures, Ed. by F. Henneberger, S. Schmitt-Rink, and E.O. Göbel, Akademie Berlin, p. 447.

Misawa, K., H. Yao, T. Hayashi, T. Kobayashi (1991): J. Chem. Phys. **94**, 4131.

Mitsunaga, M., H. Shinojima, and K. Kubodera (1988): J. Opt. Soc. Am. B **5**, 1448.

Mittleman, D.M., R.W. Schoenlein, J.J. Shiang, V.L. Colvin, A.P. Alivisatos, and C.V. Shank (1994): Phys. Rev. B **49**, 14 435.

Miyoshi, T., T. Nakatsuka, and N. Matsuo (1995): Jpn. J. Appl. Phys. **34**, 1835.

Moison, J.M., F. Houzay, F. Barthe, L. Leprince, E. Andre, and O. Vatel (1994): Appl. Phys. Lett. **64**, 196.

Müller, M.P.A., U. Lembke, U. Woggon, and I. Rückmann (1992): J. Noncryst. Solids **144**, 240.

Mui, D.S.L., D. Leonard, L.A. Coldren, and P.M. Petroff (1995): Appl. Phys. Lett. **66**, 1620.

Murray, C.B., C.R. Kagan, and M.G. Bawendi (1995): Science **270**, 1335.

Murray, C.B., D.J. Norris, and M.G. Bawendi (1993): J. Am. Chem. Soc. **115**, 8706.

Nagasawa, N., S. Koizumi, T. Mita, and M. Ueta (1976): J. Lum. **12/13**, 587.

Nagasawa, N., M. Kuwata, E. Hanamura, T. Itoh, and A. Mysyrowicz (1989): Appl. Phys. Lett. **55**, 1999.

Nair, S.V., S. Sinha, and K.C. Rustagi (1987): Phys. Rev. B **35**, 4098.

Nair, S.V. and K.C. Rustagi (1989): Superlatt. Microstruct. **6**, 377.

Naoe, K., L.G. Zimin, and Y. Masumoto (1994): Phys. Rev. B **50**, 18200.

Nash, K.J., P.D.J. Calcott, L.T. Canham, and R.J. Needs (1995): Phys. Rev. B **51**, 17698.

Neuroth, N. (1987): Optical Engineering **26**, 96.

Nirmal, M., C.B. Murray, D.J. Norris, and M.G. Bawendi (1993): Z. Phys. D **26**, 361.

Nirmal, M., D.J. Norris, M. Kuno, M.G. Bawendi, Al.L. Efros, and M. Rosen (1995): Phys. Rev. Lett. **75**, 3728.

Nötzel, R., T. Fukui, H. Hasegawa, J. Temmyo, and T. Tamamura (1994b): Appl. Phys. Lett. **65**, 2854.

Nötzel, R., N.N. Ledentsov, and K. Ploog (1993): Phys. Rev. B **47**, 1299.
Nötzel, R., J. Menninger, M. Ramsteiner, A. Ruiz, H.-P. Schönherr, and K.H. Ploog (1996): Appl. Phys. Lett. **68**, 1132.
Nötzel, R., J. Temmyo, H. Kamada, T. Furuta, and T. Tamamura (1994a): Appl. Phys. Lett. **65**, 457.
Nogami, M., K. Nagasaka, and M. Takata (1990): J. Noncryst. Solids **122**, 101.
Nojima, S. (1992): Phys. Rev. B **46**, 2302.
Nomura, S. and T. Kobayashi (1990): Sol. State Comm. **73**, 425.
Nomura, S. and T. Kobayashi (1991): Sol. State Comm. **78**, 677.
Nomura, S. and T. Kobayashi (1992a): Phys. Rev. B **45**, 1305.
Nomura, S. and T. Kobayashi (1992b): Solid State Comm. **82**, 335.
Nomura, S., and T. Kobayashi (1994): J. Appl. Phys. **75**, 382.
Nomura, S., K. Misawa, Y. Segawa, and T. Kobayashi (1993a): Phys. Rev. B **47**, 16 024.
Nomura, S., Y. Segawa, and T. Kobayashi (1993b): Sol. State. Comm. **87**, 313.
Nomura, S., Y. Segawa, and T. Kobayashi (1994): Phys. Rev. B **49**, 13571.
Norris, D.J. and M.G. Bawendi (1995): J. Chem. Phys. **103**, 5260.
Norris, D.J., M. Nirmal, C.B. Murray, A. Sacra, and M.G. Bawendi (1993): Z. Phys. D **26**, 355.
Norris, D.J., A. Sacra, C.B. Murray, and M.G. Bawendi (1994): Phys. Rev. Lett. **72**, 2612.
Nosaka, Y., K. Tanaka, N. Fujii (1993): J. Appl. Polymer Sci. **47**, 1773.
Nozue, Y. (1982): J. Phys. Soc. Japan **51**, 1840.
Nuss, M.C., W. Zinth, and W. Kaiser, (1986): Appl. Phys. Lett. **49**, 1717.

Oestreich, M., W.W. Rühle, H. Lage, D. Heitmann, and K. Ploog (1993): Phys. Rev. Lett. **70**, 1682.
Okamoto, S. and Y. Masumoto (1995): J. Lum. **64**, 253 .
Olbright, G.R., and N. Peyghambarian (1986): Appl. Phys. Lett. **48**, 1184.
Olbright, G.R., N. Peyghambarian, S.W. Koch, and L. Banyai (1987): Optics Lett. **12**, 413.
Olshavsky, M.A., A.N. Goldstein, and A.P. Alivisatos (1990): J. Am. Chem. Soc. **112**, 9438.
Onodera, Y., and Y. Toyozawa (1967): J. Phys. Soc. Jpn. **22**, 833.
Ookubo, N. (1993): J. Appl. Phys. **74**, 6375.

Pan, J., X. Xu, S. Ding, and J. Pen (1990): J. Lum. **45**, 45.
Park, S.H., R.A. Morgan, Y.Z. Hu, M. Lindberg, S.W. Koch, and N. Peyghambarian (1990): J. Opt. Soc. Am. **7**, 2097.
Pavesi, L. and M. Ceschini (1993): Phys. Rev. B **48**, 17625.
Perov, P.I., L.A. Avdeyeva, A.G. Zhdan, and N.E. Elinson (1969): Fiz. Tverd. Tela **11**, 92.
Petit, C., P. Lixon, and M.P. Pileni (1990): J. Phys. Chem. **94**, 1598.
Petterson, H., S. Anand, H. G. Grimmeiss, and L. Samuelson (1996): Phys. Rev. B **53**, R10 497.
Peyghambarian, N., B. Fluegel, D. Hulin, A. Migus, M. Joffre, A. Antonetti, S.W. Koch, and M. Lindberg (1989): IEEE J. Quantum Electronics **QE-25**, 2516.
Peyghambarian, N., H.M. Gibbs, J.L. Jewell, A. Antonetti, A. Migus, D. Hulin, and A. Mysyrowicz (1984): Phys. Rev. Lett. **53**, 2433 ; D. Hulin, A. Mysyrowicz, A. Antonetti, A. Migus, W.T. Masselink, H. Morkoc, H.M. Gibbs, and N. Peyghambarian (1986): Phys. Rev. B **32**, 4389.

Peyghambarian, N., S.W. Koch, and A. Mysyrowicz (1993): Introduction to Semi-conductor Optics, Prentice Hall.

Pfannkuche, D. and S. Ulloa (1995): Phys. Rev. Lett. **74**, 1194.

Pickering, C., M.I.J. Beale, D.J. Robbins, P.J. Pearson, and R. Greef (1984): J.Phys. **C17**, 6535.

Pollock, E.L. and S.W. Koch (1991): J. Chem. Phys. **94**, 6766.

Potter B.G. and J.H. Simmons (1988): Phys. Rev. B **37**, 10838.

Priester, C. and M. Lannoo (1995): Phys. Rev. Lett. **75**, 93.

Prokes, S.M. (1993): J. Appl. Phys. **73**, 407.

Prokes, S.M., W.E. Carlos, and O.J. Glembocki (1994): Phys. Rev. B **50**, 17093.

Proot, J.P., C. Delerue, and G. Allan (1992): Appl. Phys. Lett. **61**, 1948.

Ramaniah, L.M. and S.V. Nair (1993): Phys. Rev. B **47**, 7132.

Ramaniah, L.M., S.V. Nair, and K.C. Rustagi (1989): Phys. Rev. B **40**, 12423.

Reed, M.A. (1993): Scientific American **268**, 98.

Reed, M.A. and W. Kirk (eds.) (1991): Nanostructures and Mesoscopic Systems, Academic, San Diego.

Reisfeld, R. (1992): NATO ASI Ser. **B 301**, 19.

Remillard, J.T., H. Wang, M.D. Webb, and D.G. Steel (1989): IEEE J. Quantum Electr. **25**, 408.

Ricolleau, C., L. Audinet, M. Gandais, T. Gacoin, J.P. Boilot, and M. Chamarro (1996): J. Cryst. Growth **159**, 861.

Roca, E., C. Trallero-Giner, and M. Cardona (1994): Phys. Rev. B **49**, 13704.

Rocksby, H.P. (1932): J. Soc. Glass Techn. **16**, 171.

Rodden, W.S.O., C.N. Ironside, and C.M. Sotomayor Torres (1994): Semicond. Sci. Technol. **9**, 1839.

Rodden, W.S.O., C.M. Sotomayor Torres, and C.N. Ironside (1995): Semicond. Sci. Technol. **10**, 807.

Rodrigues, P.A.M., G. Tamulaitis, P.Y. Yu, and S.H. Risbud (1995): Sol. State Comm. **94**, 583.

Romestain, R. and G. Fishman (1994): Phys. Rev. B **49**, 1774.

Rorison, J.M. (1993): Phys. Rev. B **48**, 4643.

Rosenbauer, M., S. Finkbeiner, E. Bustarret, J. Weber, and M. Stutzmann (1995): Phys. Rev. B **51**, 10539.

Rosetti, R., R. Hull, J.L. Ellison, and L.E. Brus (1984): J. Chem. Phys. **80**, 4464.

Rosetti, R., S. Nakahara, and L.E. Brus (1983): J. Chem. Phys. **79**, 1086.

Rössler, U. and H.R. Trebbin (1981): Phys. Rev. B **23**, 1961.

Rossmann, H., A. Schülzgen, F. Henneberger, and M. Müller (1990): phys. stat. sol. (b) **159**, 287.

Rotter, P., U. Rössler, H. Silberbauer, and M. Suhrke (1995): Physica B **212**, 231.

Rougemont, F. de, R. Frey, P. Roussignol, D. Ricard, and C. Flytzanis (1987): Appl. Phys. Lett. **50**, 1619.

Roussignol, P., M. Kull, D. Ricard, F. de Rougemont, R. Frey, and C. Flytzanis (1987b): Appl. Phys. Lett. **51**, 1882; erratum (1989) **54**, 1705.

Roussignol, P., D. Ricard, J. Lukasik, and C. Flytzanis (1987a): J. Opt. Soc. Am. B **4**, 5.

Roussignol, P., D. Ricard, C. Flytzanis, and N. Neuroth (1989): Phys. Rev. Letters **62**, 312.

Roussignol, P., D. Ricard, K.C. Rustagi, and C. Flytzanis (1985): Opt. Comm. **55**, 143.

Roy, A., and A.K. Sood (1996): Solid State Communications **97**, 97.

Ruppin, R. (1975): J. Phys. C **8**, 1969.

Rustagi, K.C. and C. Flytzanis (1984): Optics Lett. **9**, 344.

Ruvimov, S., P. Werner, K. Scheerschmidt, U. Gösele, J. Heydenreich, U. Richter, N.N. Ledentsov, M. Grundmann, D. Bimberg, V.M. Ustinov, A.Yu. Egorov, P.S. Kopev, and Zh. I. Alferov (1995): Phys. Rev. B **51**, 14766.

Sacra, A., D.J. Norris, C.B. Murray, and M.G. Bawendi (1995): J. Chem. Phys. **103**, 5236.

Salata, O.V., P. Dobson, P.J. Hull, and J.L. Hutchinson (1994): Appl. Phys. Lett. **65**, 189.

Saltiel, S.M., B. van Wonterghem, and P.M. Rentzepis (1990): Opt. Comm. **77**, 59.

Scamarcio, G., M. Lugara, and D. Manno (1992): Phys. Rev. B **45**, 13792.

Schanne-Klein, M.C., F. Hache, D. Ricard, and C. Flytzanis (1992): J. Opt. Soc. Am. B **9**, 2234.

Schmidt, H.M. and H. Weller (1986): Chem. Phys. Lett. **129**, 615.

Schmitt-Rink, S., D.S. Chemla, and D.A.B. Miller (1985): Phys. Rev. B **32**, 6601.

Schmitt-Rink, S., D.A.B. Miller, and D.S. Chemla (1987): Phys. Rev. B **35**, 8113.

Schoenlein, R.W., D.M. Mittleman, J.J. Shiang, A.P. Alivisatos, and C.V. Shank (1993): Phys. Rev. Lett. **70**, 1014.

Scholze, H. (1977): Glas – Natur, Struktur und Eigenschaften, 2nd edition, Springer, Berlin.

Schooss, D., A. Mews, A. Eychmüller, and H. Weller (1994): Phys. Rev. B **49**, 17072.

Schubert, M. and B. Wilhelmi (1978): Einführung in die Nichtlineare Optik/T.II, BSB Teubner Leipzig.

Schuppler, S., S.L. Friedman, M.A. Marcus, D.L. Adler, Y.H. Xie, F.M. Ross, T.D. Harris, W.L. Brown, Y.J. Chabal, L.E. Brus, and P.H. Citrin (1994): Phys. Rev. Lett. **72**, 2648.

Schwab, H., C. Dörnfeld, E.O. Göbel, J. Hvam, C. Klingshirn, J. Kuhl, V.G. Lyssenko, F.A. Majumdar, G. Noll, J. Nunnenkamp, K.H. Pantke, A. Resnitzy, U. Siegner, H.E. Swoboda, Ch. Weber (1992): phys. stat. sol. (b) **172**, 479.

Sekikawa, T., H.Yao, T.H.Hayashi, and T.Kobayashi (1992): Sol. State Comm. **83**, 969.

Sercel, P. (1995): Phys. Rev. B **51**, 14532.

Sercel, P.C. and K.J. Vahala (1990): Phys. Rev. B **42**, 3690.

Shah, J., B. Deveaud, T. Damen, W. Tsang, A. Gossard, and P. Luigli (1987): Phys. Rev. Lett. **59**, 2222.

Shen, Y. (1984): The Principles of Nonlinear Optics, John Wiley, New York.

Shepilov, M.P. (1992): J. Noncryst. Solids **146**, 1.

Shiang, J.J., S.H. Risbud, and A.P. Alivisatos (1993): J. Chem. Phys. **98**, 8432.

Shinojima, H., J. Yumoto, and N. Uesugi (1992): Appl. Phys. Lett. **60**, 298.

Siegman, A.E. (1977): Appl. Phys. Lett. **30**, 21.

Smith, D.L. and C. Mailhiot (1990): Rev. Mod. Phys. **62**, 173.

Smith, R.L. and S.D. Collins (1992): J. Appl. Phys. **71**, R1.

Smith, S.D. (1984): Acta Physica Austriaca **56**, 75.

Snavely, B.B. (1968): Phys. Rev. **167**, 730.

Solimany, L. and B. Kramer (1995): Solid State Comm. **96**, 471.

Sommerfeld, A. (1988): Vorlesungen über theoretische Physik, Band III (Elektrodynamik): Kapitel 11, Verlag Harry Deutsch, Thun, Frankfurt/M. .

Sotomayor Torres, C.M., A.P. Smart, M.A. Foad, and C.D.W. Wilkinson (1992): Advances in Solid State Physics (Festkörperprobleme) **32**, 265.

Sotomayor Torres, C.M., A.P. Smart, M. Watt, M.A. Foad, K. Tsutsui, and C.D. Wilkinson (1994): J. Electronic Materials **23**, 289.

Spanhel, L. and M. Anderson (1991): J. Am. Chem. Soc. **112**, 2278 (1990); **113**, 2826 (1991).

Spanhel, L., E. Arpac, and H. Schmidt (1992a): J. Noncryst. Solids **147/148**, 657.

Spanhel, L., H. Haase, H. Weller, and A. Henglein (1987a): J. Amer. Chem. Soc. **109**, 5649.

Spanhel, L., H. Schmidt, A. Uhrig, and C. Klingshirn (1992b): Proc. MRS Spring Meeting, San Francisco 1992.

Spanhel, L., H. Weller, and A. Henglein (1987b): J. Am. Chem. Soc. **109**, 6632.

Sperling, V., U. Woggon, A. Lohde, and T. Haalboom (1996): Int. Conf. Luminescence, Prag 1996, to be published in J. Lum.; A. Lohde, Diploma Thesis, University of Karlsruhe (1996).

Steffen, R., Th. Koch, J. Oshinowo, F. Faller, and A. Forchel (1996): Appl. Phys. Lett. **68**, 223.

Stegeman, G.I., and R.H. Stolen (1989): J. Opt. Soc. Am. B **6**, 652.

Steigerwald, M.L., A.P. Alivisatos, J.M. Gibson, T.D. Harris, R. Kortan, A.J. Muller, A.M. Thayer, T.M. Duncan, D.C. Douglass, and L.E. Brus (1988): J. Amer. Chem. Soc. **110**, 3046.

Steigerwald, M.L., and L.E. Brus (1989): Annual Rev. Mater. Sci. **19**, 471.

Stephani, H., and G. Kluge (1995): Theoretische Mechanik, Spektrum Akademischer Verlag, Heidelberg, Berlin, Oxford.

Stranski, I.N., and L. von Krastanow (1939): Akad. Wiss. Lit. Mainz, Abh. Math. Naturwiss. Kl. IIb **146**, 797.

Stutzmann, M., M.S. Brandt, M. Rosenbauer, J. Weber, and H.D. Fuchs (1993): Phys. Rev. B **47**, 4806.

Stutzmann, M., J. Weber, M.S. Brandt, H.D. Fuchs, M. Rosenbauer, P. Deak, A. Höpner, and A. Breitschwerdt (1992): Adv. Solid State Phys. **32**, 179.

Suemoto, T., K. Tanaka, and A. Nakajima, Phys. Rev. B **49**, 11005.

Suemoto, T., K. Tanaka, A. Nakajima, and T. Itakura (1993): Phys. Rev. Lett. **70**, 3659.

Suga, S., T. Koda, and T. Mitani (1971): Phys. Status Solidi **48**, 753.

Sugiyama, Y., Y. Sakuma. S. Muto, and N. Yokoyama (1995): Appl. Phys. Lett. **67**, 256.

Takagahara, T. (1989): Phys. Rev. B **39**, 10206.

Takagahara, T. (1993a): Phys. Rev. B **47**, 4569.

Takagahara, T. (1993b): Phys. Rev. Lett. **71**, 3577.

Takagahara, T. and K. Takeda (1992): Phys. Rev. B **46**, 15578.

Takagahara, T. and K. Takeda (1996): Phys. Rev. B **53**, R4205.

Takagi, H., H. Ogawa, Y. Yamazaki, A. Ishizaki, and T. Nakagiri (1990): Appl. Phys. Lett. **56**, 2379.

Tamura, A., K. Higeta, and T. Ichinokawa (1982): J. Phys. C: Solid State Phys. **15**, 4975.

Tamura, H., M. Rückscloss, T. Wirschem, and S. Veprek (1994): Appl. Phys. Lett. **65**, 1537.

Tanaka, A., S. Onari, and T. Arai, (1992): Phys. Rev. B **45**, 6587.

Tanaka, A., S. Onari, and T. Arai, (1993): Phys. Rev. B **47**, 1237.

Tanaka, Y. and H. Akera (1996): Phys. Rev. B **53**, 3901.

Tillmann, K., D. Gerthsen, P. Pfundstein, A. Förster, and K. Urban (1995): J. Appl. Phys. **78**, 3824.

Tokizaki, T., H. Akiyama, M. Takaya, and A. Nakamura (1992): J. Cryst. Growth **117**, 603.

Tokizaki, T., Y. Ishida, and T. Yajima (1989): Opt. Comm. **71**, 355.

Tolbert, S.H., and A.P. Alivisatos (1994b): Science **265**, 373.

Tolbert, S.H., A.B. Herold, C. S. Johnson, and A.P. Alivisatos (1994a): Phys. Rev. Lett. **73**, 3266.

Tomita, M. and M. Matsuoka (1990): J. Opt. Soc. Am. B **7**, 1198.

Tran Thoai, D.B., Y.Z. Hu, and S.W. Koch (1990): Phys. Rev. B **42**, 11261.

Trinn, M., S. Freundt, U. Woggon, M. Müller, and R.G. Ulbrich (1991): Spring Meeting of the German Physical Society (Technical Digest), Münster 1991.

Tsekhomskii, V.A. (1978): Phys. Chem. Glass **4**, 3.

Tsybeskov, L. and P.M. Fauchet (1994a): Appl. Phys. Lett. **64**, 1983.

Tsybeskov, L., Ju. V. Vandyshev, and P.M. Fauchet (1994b): Phys. Rev. B **49**, 7821.

Tulkki, J. and A. Heinämäki (1995): Phys. Rev. B **52**, 8239.

Turnbull, D. (1956): Solid State Phys. **3**, 255.

Uhrig, A., L. Banyai, S. Gaponenko, A. Wörner, N. Neuroth, and C. Klinghsirn (1991): Z. Phys. D **20**, 345.

Uhrig, A., L. Banyai, Y.Z. Hu, S.W. Koch, C. Klingshirn, and N. Neuroth (1990): Z. Phys. B **81**, 385.

Uhrig, A. (1992): A. Wörner, C. Klingshirn, L. Banyai, S. Gaponenko, I. Lacis, N. Neuroth, B. Speith, and K. Remitz (1992): J. Cryst. Growth **117**, 598.

Vandyshev, Yu.V., V.S. Dneprovskii, and V.I. Klimov (1992): Sov. Phys. - JETP **74**, 144.

Van Houten, H. (1992): Surface Science **263**, 442.

Van Wonterghem, B., S.M. Saltiel, T.E. Dutton, and P.M. Rentzepis (1989): J. Appl. Phys. **66**, 4935.

Vasil'ev, M.I., N.A.Grigor'ev, B.S.Kulinkin, and V.A.Tsekhomskii (1991): Phys. Chem. Glass **17**, 594.

Vasques, R.P., R.W.Fathauer, T.George, A.Ksendsov, and T.L.Lin (1992): Appl. Phys. Lett. **60**, 1004.

Vial, J.C., A. Bsiesy, F. Gaspard, R. Herino, M. Ligeon, F. Muller, R. Romestain, and R.M. Macfarlane (1992): Phys. Rev. B **45**, 14171.

Vossmeyer, T., G. Reck, B. Schulz, L. Katsikas, and H. Weller (1995): J. Am. Chem. Soc. **117**, 12881.

Wamura, T., Y. Masumoto, and T. Kawamura (1991): Appl. Phys. Lett. **59**, 1758.

Wang, Lin-Wang and A. Zunger (1996): Phys. Rev. B **53**, 9579.

Wang, P.D., C. Sotomayor Torres, H. Benisty, C. Weisbuch, and S.P. Beaumont (1992): Appl. Phys. Lett. **61**, 946.

Wang, Y. (1991a): Acc. Chem. Res. **24**, 133.

Wang, Y. (1995): Adv. Photochem. **19**, 179.

Wang, Y., and N. Herron (1990): Phys. Rev. B **42**, 7253.

Wang, Y., and N. Herron (1991b): J. Phys. Chem. **95**, 525.

Wang, Y., and N. Herron (1992b): Chem. Phys. Lett. **200**, 71.

Wang, Y., N. Herron, W. Mahler, A. Suna (1989): J. Opt. Soc. Am. B **6**, 808.

Wassermeier, M., J. Sujiono, M.D. Johnson, K.T. Leung, B.G. Orr, L. Däweritz, and K. Ploog (1995): Phys. Rev. B **51**, 14721.

Weller, H. (1991): Ber. Bunsenges. Phys. Chem. **95**, 1361.

Weller, H., U. Koch, M. Gutierrez, and A. Henglein (1984): Ber. Bunsenges. Phys. Chem. **88**, 649.

Weller, H., H.M. Schmidt, U. Koch, A. Fojtik, S. Baral, A. Henglein, W. Kunath, K. Weiss, and E. Dieman (1986): Chem. Phys. Lett. **129**, 615.

Wind, O., F. Gindele, U. Woggon, and C. Klingshirn (1996): J. Cryst. Growth **159**, 867.

Wind, O., H. Kalt, U. Woggon, and C. Klingshirn (1994): Adv. Mat. Electr. Opt. **3**, 89.

Wörner, A. (1992): Diploma Thesis, University Kaiserslautern, Germany.

Woggon, U. (1995a): In: Nonlinear Spectroscopy of Solids: Advances and Applications, ed. B. Di Bartolo, NATO ASI series B, Plenum, Vol. **339**, 425.

Woggon, U. (1996a): In: Adv. in Solid State Physics (Festkörperprobleme) **35**, ed. R. Helbig, Vieweg, p. 145.

Woggon, U., S.V. Bogdanov, O. Wind, K.-H. Schlaad, H. Pier, C. Klingshirn, P. Chatziagorastou, and H.-P. Fritz (1993a): Phys. Rev. B **48**, 11979.

Woggon, U., S.V. Bogdanov, O. Wind, and V. Sperling (1994a): J. Cryst. Growth **138**, 976.

Woggon, U. and S.V. Gaponenko (1995b): physica status solidi (b) **189**, 285.

Woggon, U., S. Gaponenko, W. Langbein, A. Uhrig, and C. Klingshirn (1993b): Phys. Rev. B **47**, 3684.

Woggon, U., S.V. Gaponenko, A. Uhrig, W. Langbein, and C. Klingshirn (1994b): Adv. Mat. Opt. El. **3**, 151.

Woggon, U., H. Giessen, F. Gindele, O. Wind, B. Fluegel, and N. Peyghambarian (1996d): to be published

Woggon, U., F. Gindele, O. Wind, and C. Klingshirn (1996c): Phys. Rev. B **54**, 1506.

Woggon, U., F. Henneberger, and M. Müller (1988): phys. stat. sol.(b) **150**, 641.

Woggon, U., M. Müller, U. Lembke, I. Rückmann, J. Cesnulevicius (1991a): Superlatt. Microstruct. **9**, 245.

Woggon, U., M. Müller, I. Rückmann, J. Kolenda, and M. Petrauskas (1990): phys. stat. sol. (b) **160**, K79.

Woggon, U. and M. Portuné (1995c): Phys. Rev. B **51**, 4719.

Woggon, U., M. Portuné, C. Klingshirn, H. Giessen, B. Fluegel, G. Mohs, and N. Peyghambarian (1995d): phys. stat. sol. (b) **188**, 221.

Woggon, U., I. Rückmann, J. Kornack, M. Müller, J. Cesnulevicius, J. Kolenda, and M. Petrauskas (1991b): SPIE Proc. **1362**, 885.

Woggon, U., I. Rückmann, J. Kornack, M. Müller, J. Cesnulevicius, M. Petrauskas, and J. Kolenda (1992): J. Cryst. Growth **117**, 608.

Woggon, U., M. Saleh, A. Uhrig, M. Portuné, and C. Klingshirn (1994c): J. Cryst. Growth **138**, 988; Sperling, V., Diploma Work, University of Kaiserslautern 1994.

Woggon, U., O. Wind, F. Gindele, E. Tsitsishvili, and M. Müller (1996b): J. Lum., to be published; F. Gindele, Diploma thesis, University Karlsruhe (1996).

Woggon, U., O. Wind, W. Langbein, O. Gogolin, and C. Klingshirn (1994d): J. Lum. **59**, 135.

Woggon, U., O. Wind, W. Langbein, and C. Klingshirn (1995e): Jpn. J. Appl. Phys. **34**, Suppl. 34-1, 232.

Xia, Jian-Bai (1989): Phys. Rev. B **40**, 8500.

Xie, Y.H., M.S. Hybertsen, W.L. Wilson, S.A. Ipri, G.E. Carver, W.L. Brown, E. Dons, B.E. Weir, A.R. Kortan, G.P. Watson, and A.J. Liddle (1994): Phys. Rev. B **49**, 5386.

Xie, Y.H., W.L. Wilson, F.M. Ross, J.A. Mucha, E.A. Fitzgerald, J.M. Maccaulay, and T.D. Harris (1992): J. Appl. Phys. **71**, 2403.

Xu, Z.Y., Z.L. Yuan, J.Z. Xu, B.Z. Zheng, B.S. Wang, D.S. Jiang, R. Nötzel, and K. Ploog (1995): Phys. Rev. B **51**, 7024.

Yajima, T., and H. Souma (1978): Phys. Rev. A **17**, 309.
Yajima, T., H. Souma, and Y. Ishida (1978): Phys. Rev. A **17**, 324.
Yamamoto, M., R. Hayashi, K. Tsunetamo, K. Kohno, and Y.Osaka (1991): Jpn. J. Appl. Phys. **30**, 136.
Yan, F. and J.M. Parker (1989): J. Noncryst. Solids **112**, 277.
Yanagida, S., T. Enokida, A. Shindo, T. Shiragami, T. Ogata, T. Fukumi, T. Sakaguchi, H. Mori, and T. Sakata (1990): Chemistry Lett., 1773.
Yao, S.S., C. Karaguleff, R. Fortenberry, C.T. Seaton, and G.I. Stegeman (1985): Appl. Phys. Lett. **46**, 801.
Yariv, A. (1989): Quantum Electronics, 3rd. Ed., John Wiley, New York.
Yumoto, J., S. Fukushima, and K. Kubodera (1987): Opt. Lett. **12**, 832.
Yükselici, H., P.D. Persans, and T.M. Hayes (1995): Phys. Rev. B **52**, 11763.

Zhang, Y., M. Sturge, K. Kash, B.P. van der Gaag, A.S. Gozdz, L.T. Florez, and J.P. Harbison (1995): Phys. Rev. B **51**, 13303.
Zimin, L.G., S.V. Gaponenko, V.Yu. Lebed, I.E. Malinovskii, and I.N. Germanenko (1990): J. Lum. **46**, 101.
Zrenner, A., L.V. Butov, M. Hagn, G. Abstreiter, G. Böhn, and G. Weimann (1994): Phys. Rev. Lett. **72**, 3382.
Zou, J., D.J.H. Cockayne, and B.F. Usher (1996): Appl. Phys. Lett. **68**, 673.

Index

Printing: Mercedesdruck, Berlin
Binding: Buchbinderei Lüderitz & Bauer, Berlin

Springer Tracts in Modern Physics